Chris Killip
Pirelli Work

With an essay by Clive Dilnot

Chris Killip Pirelli Work

Steidl

In Memoriam

Tom Northey
1938 – 1996

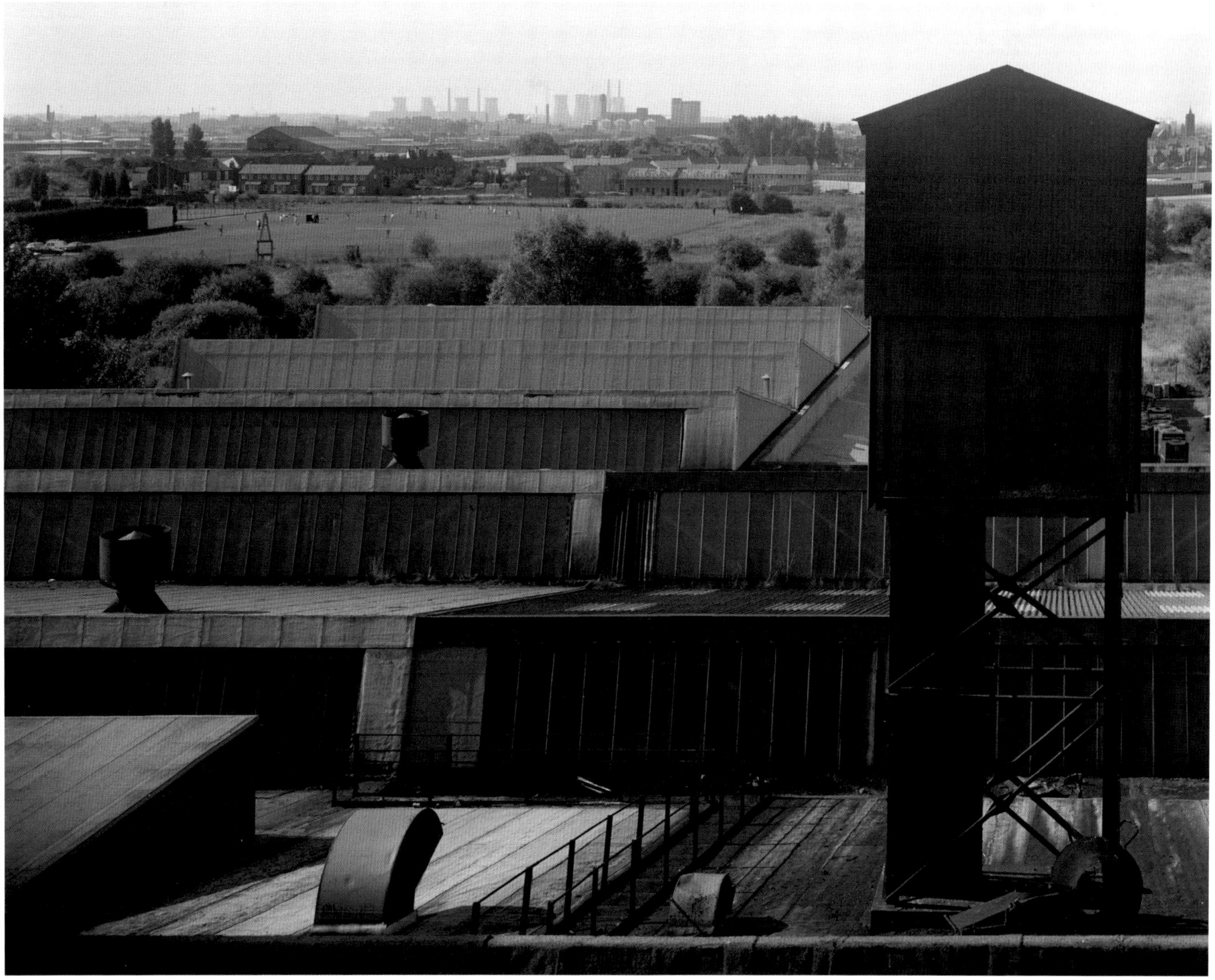

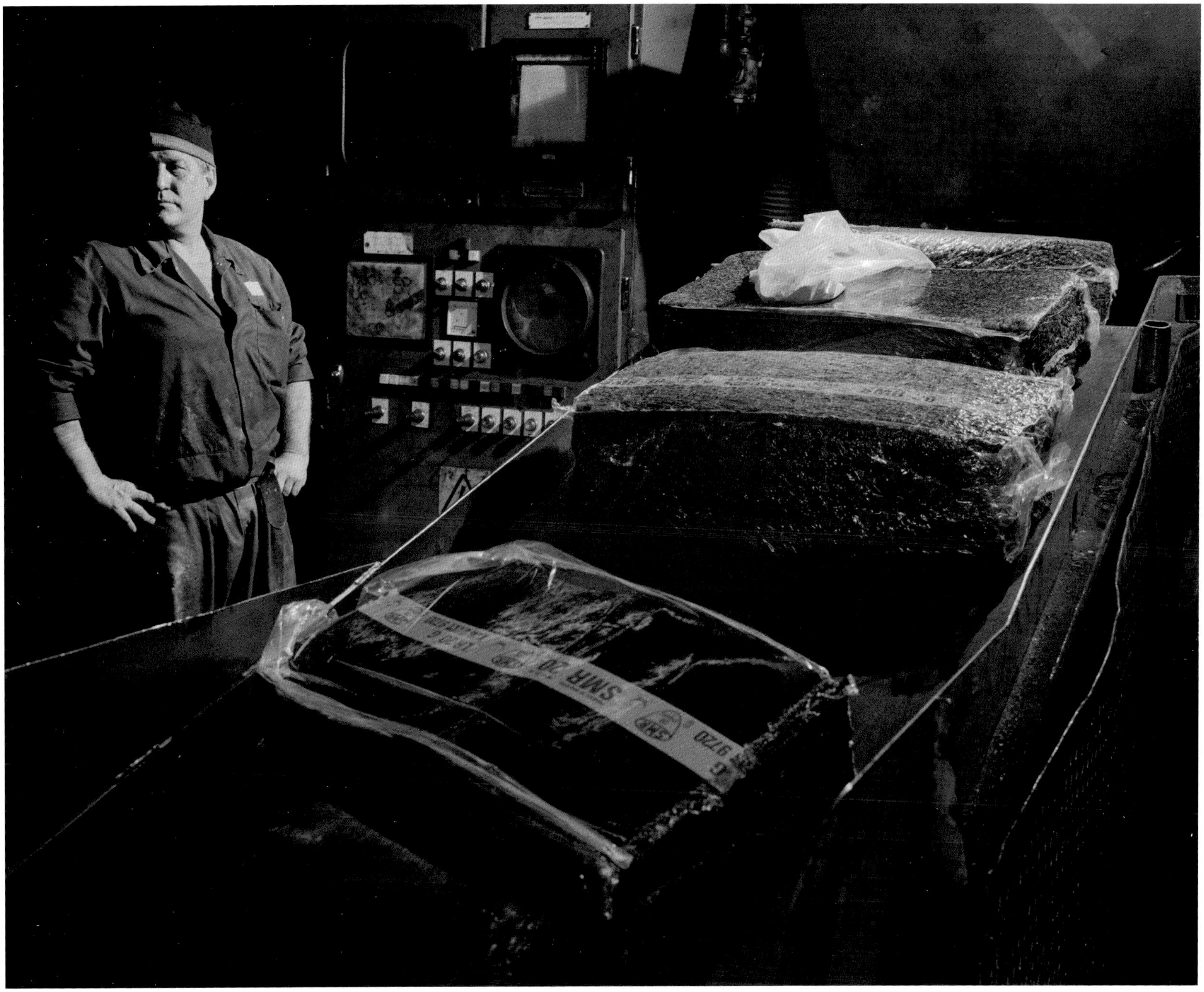
SMR 20

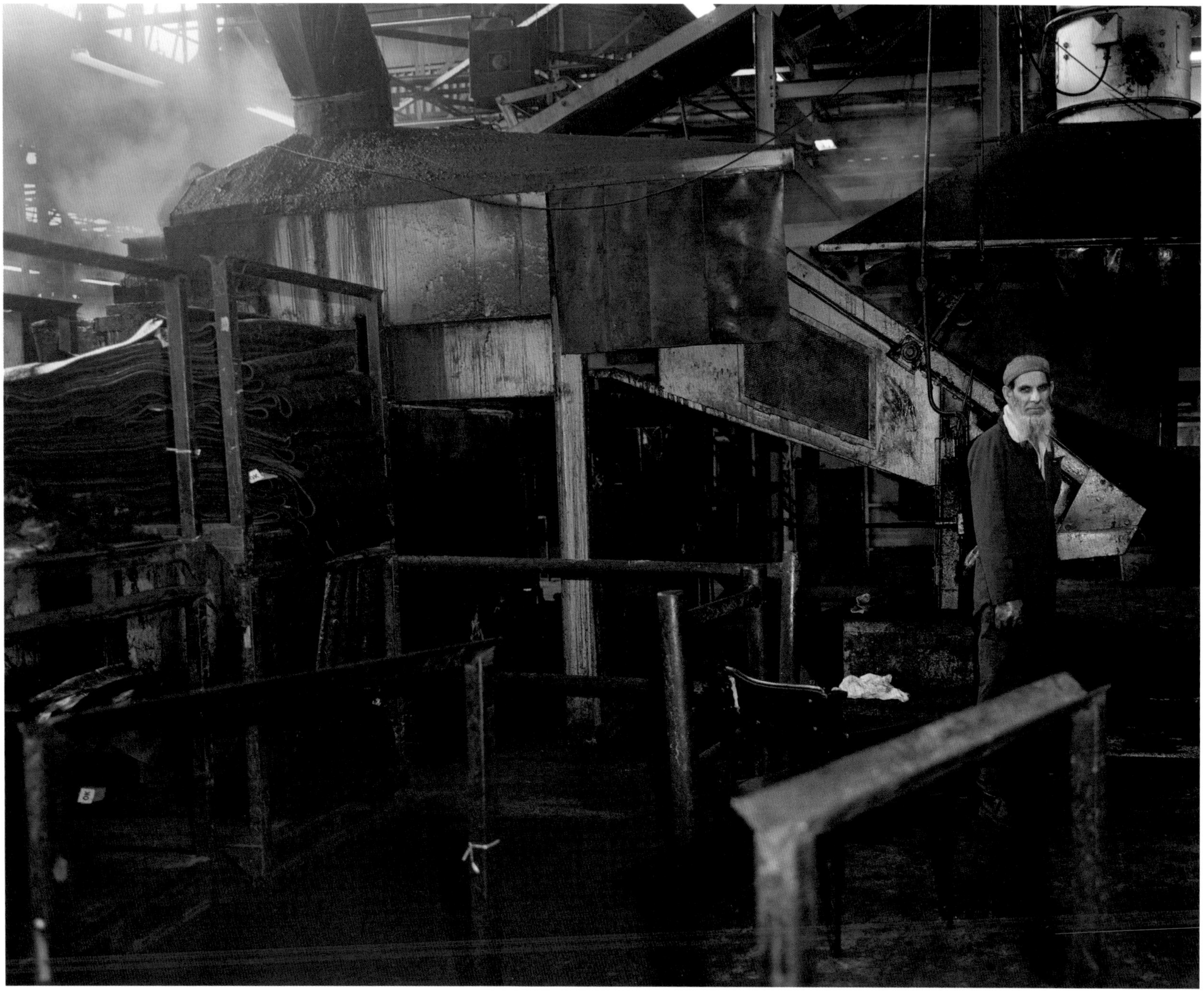

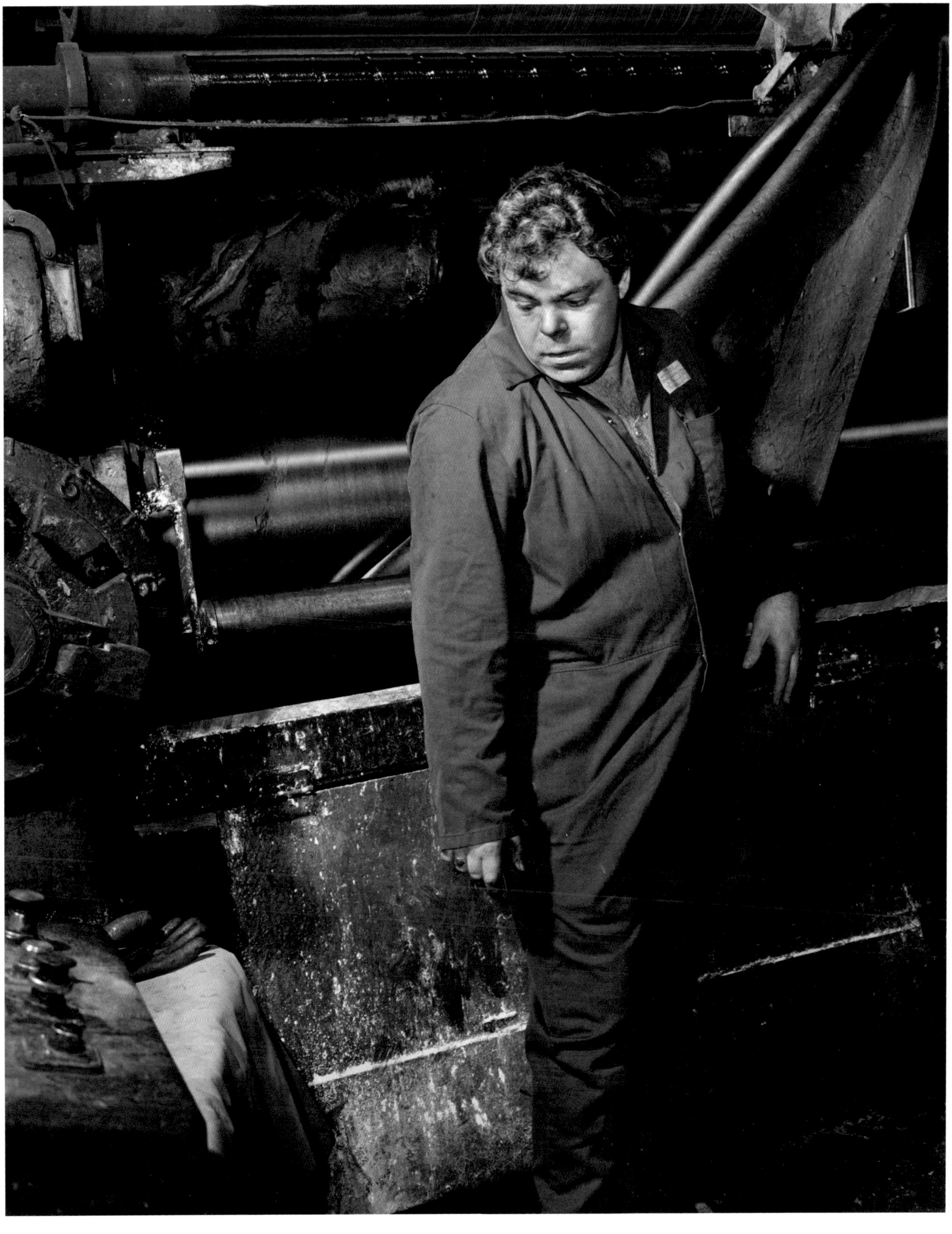

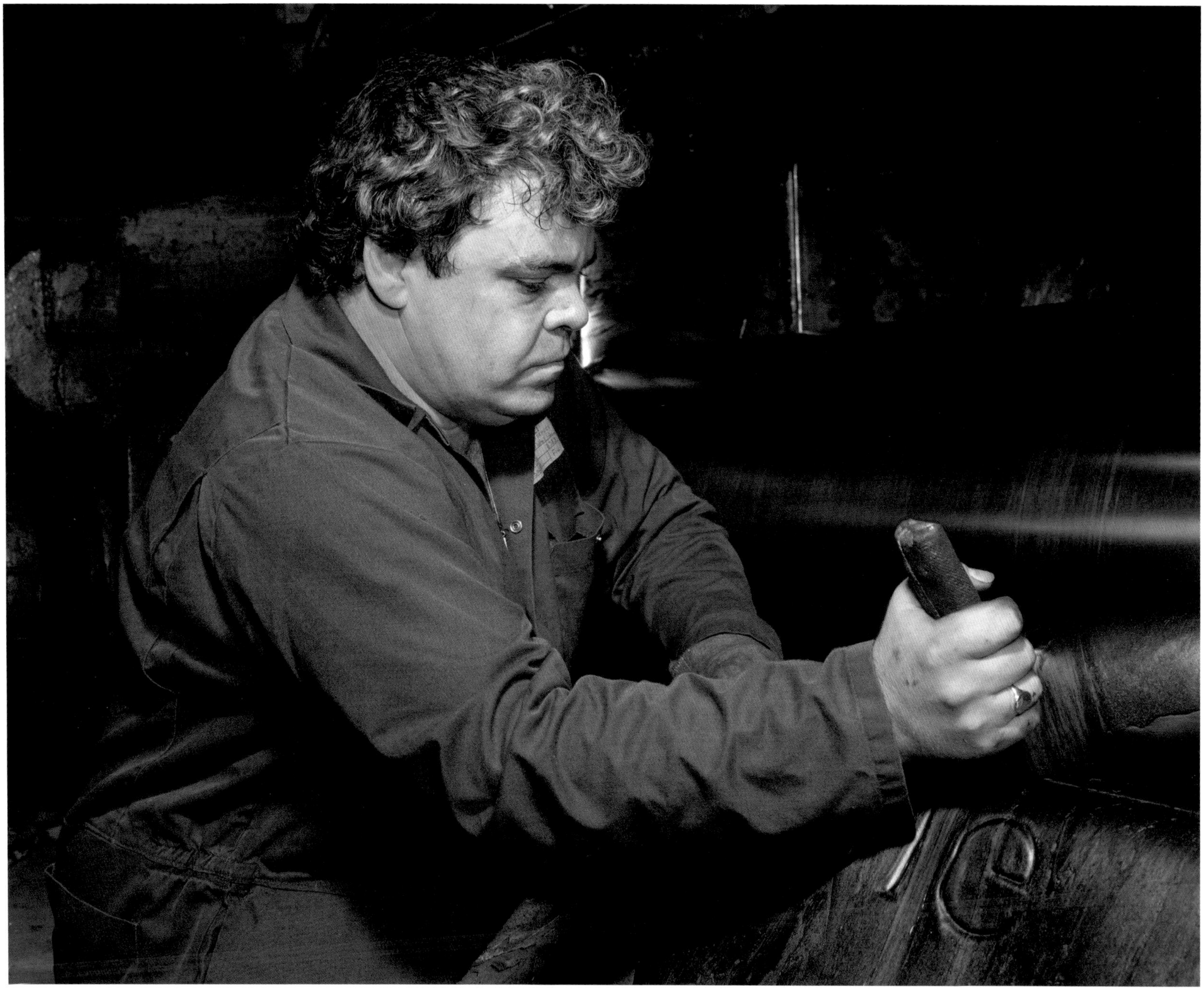

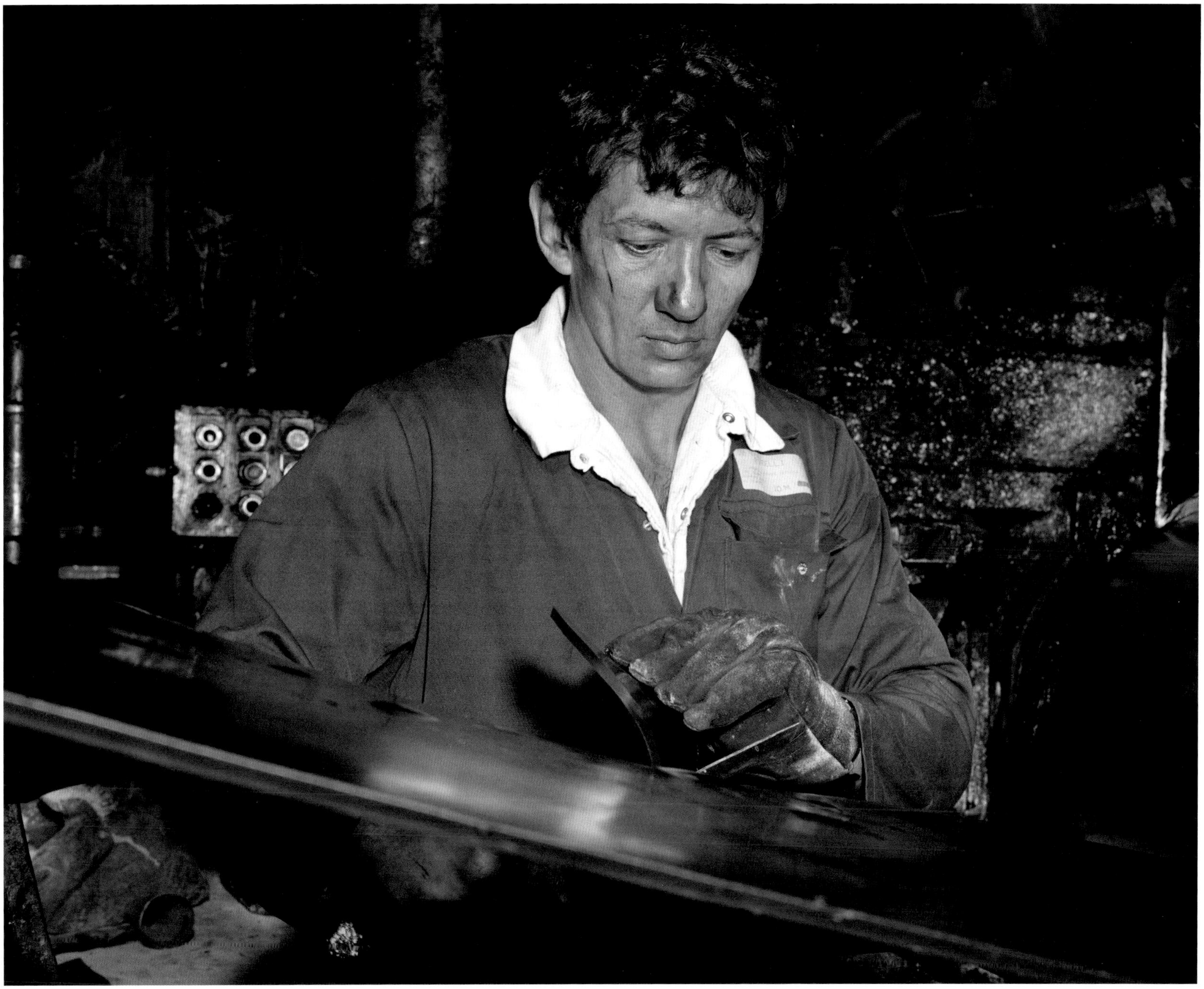

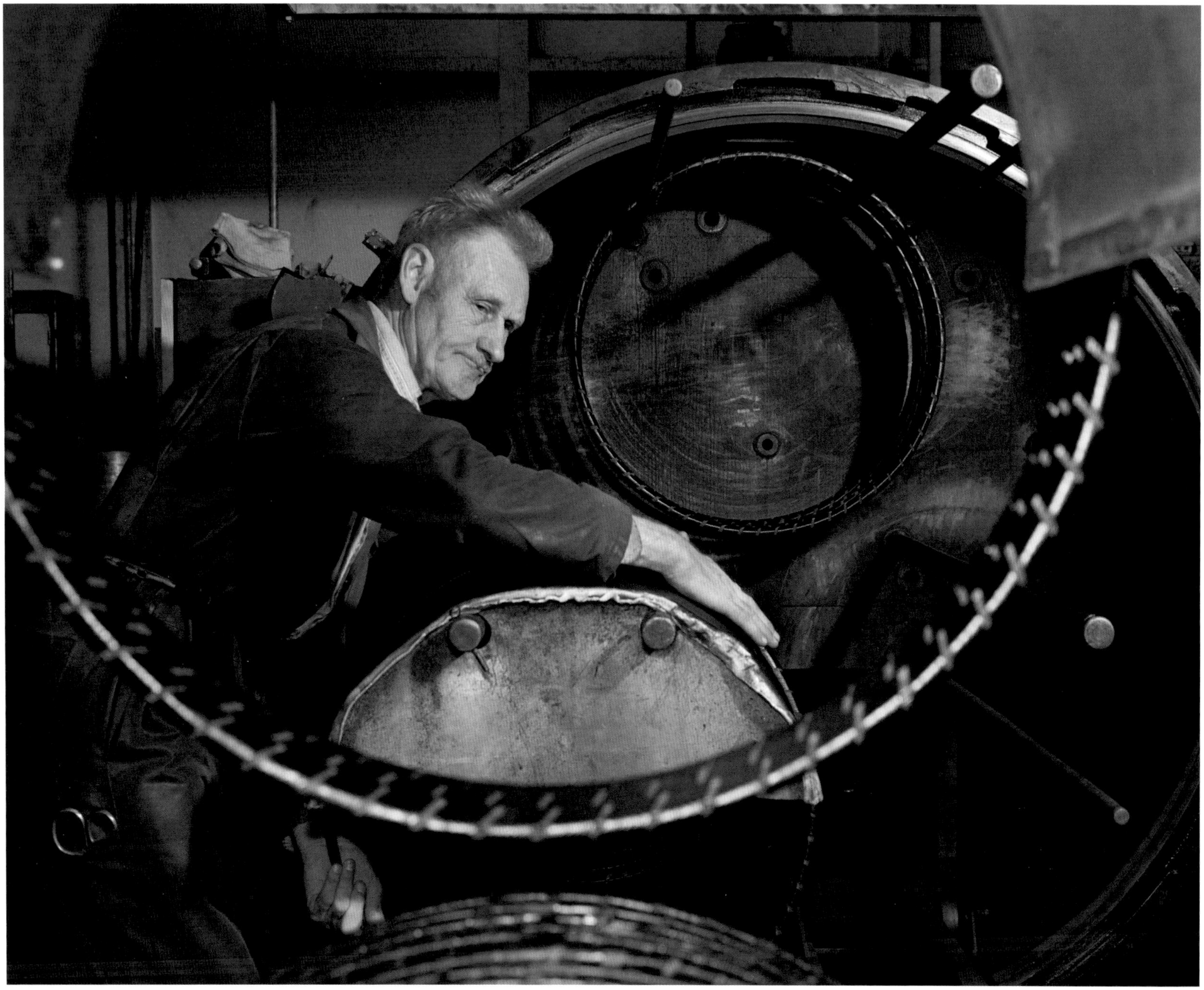

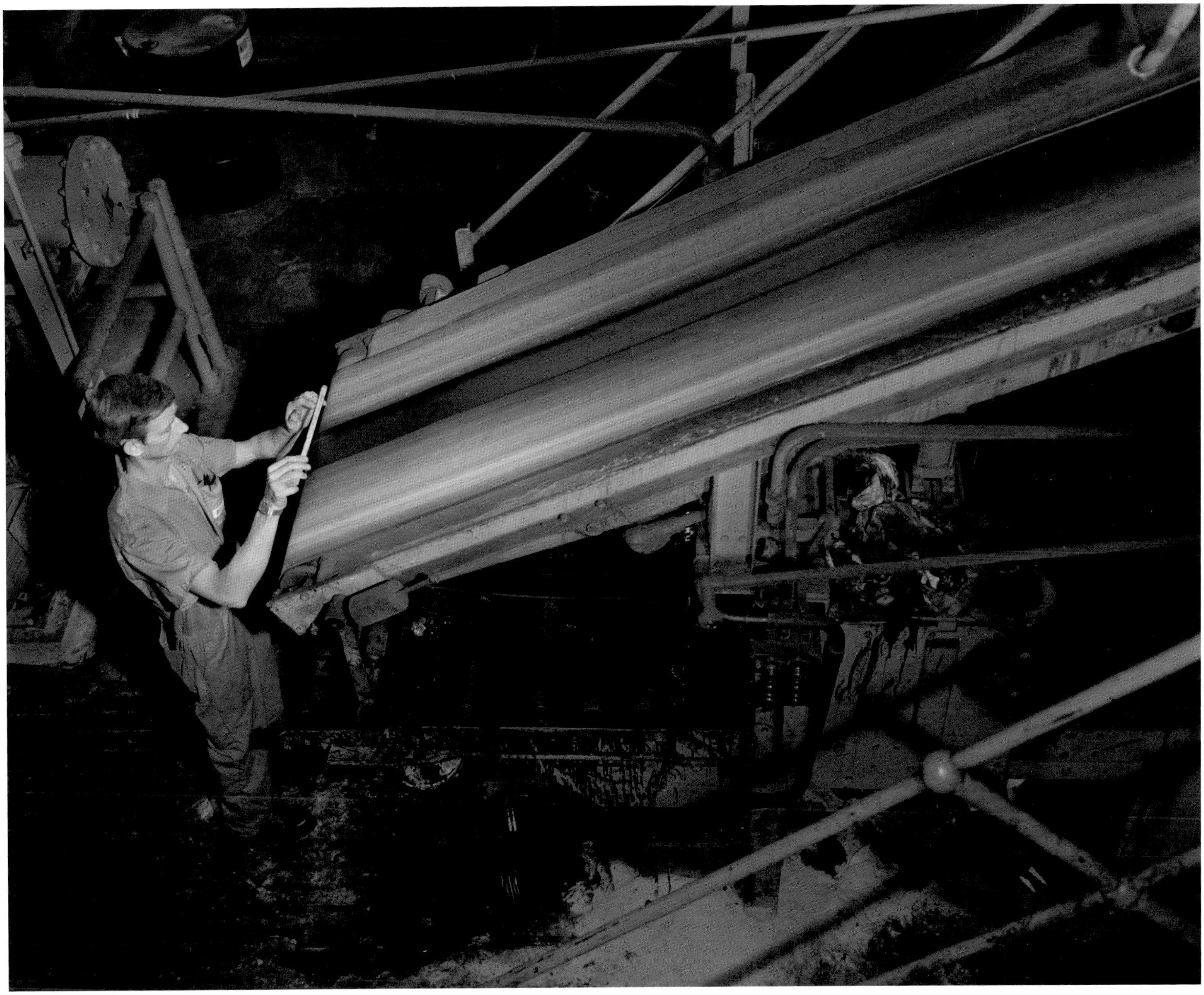

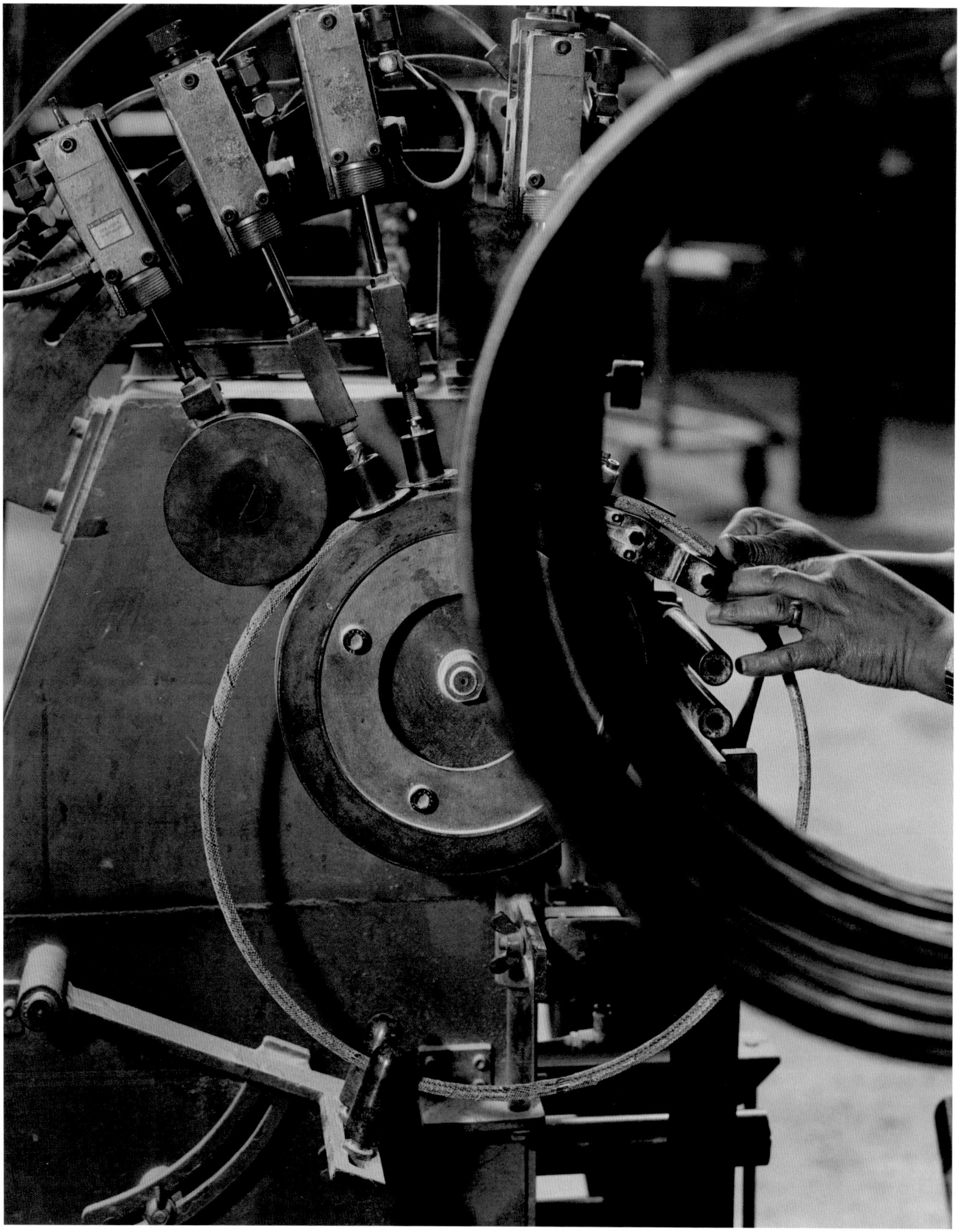

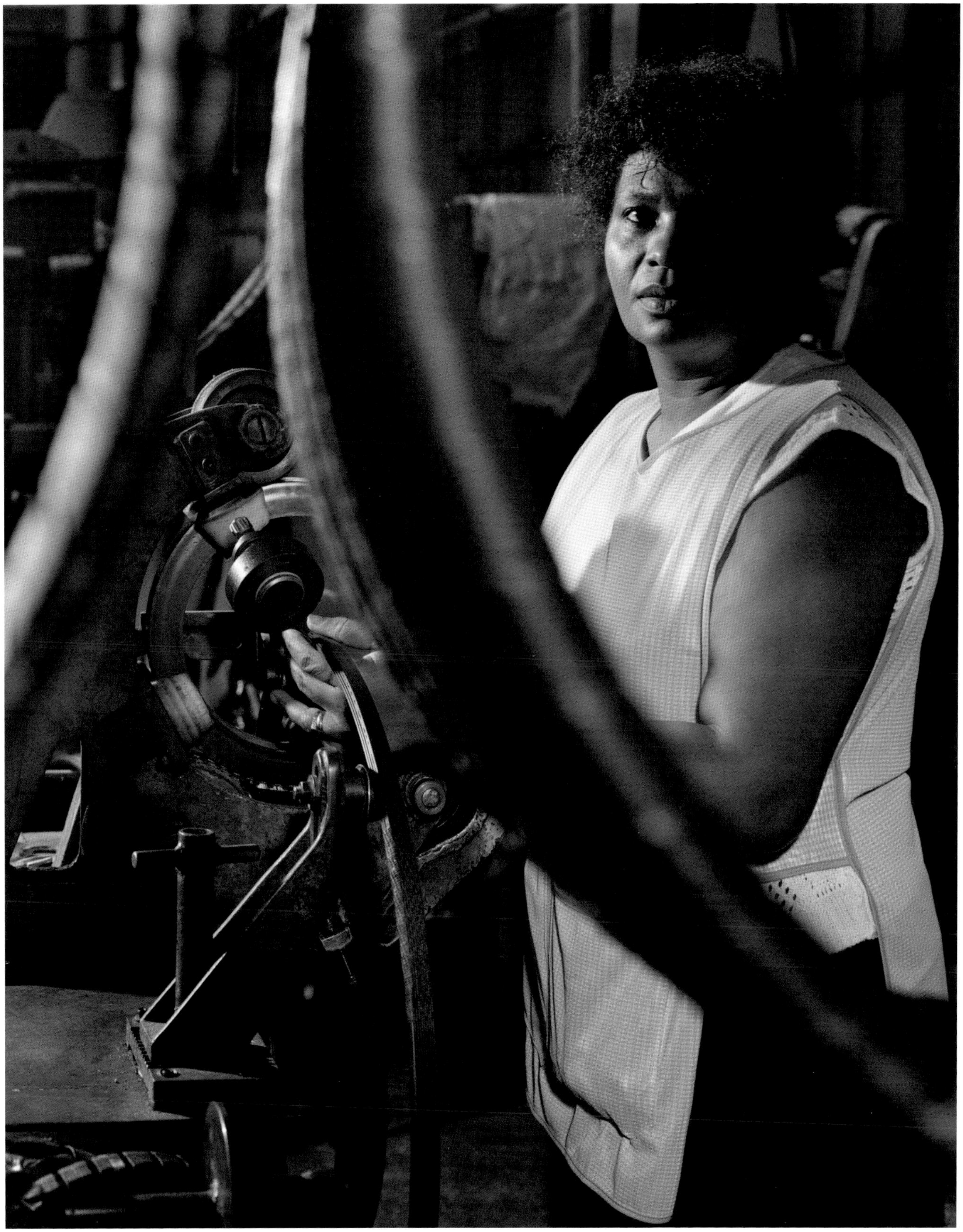

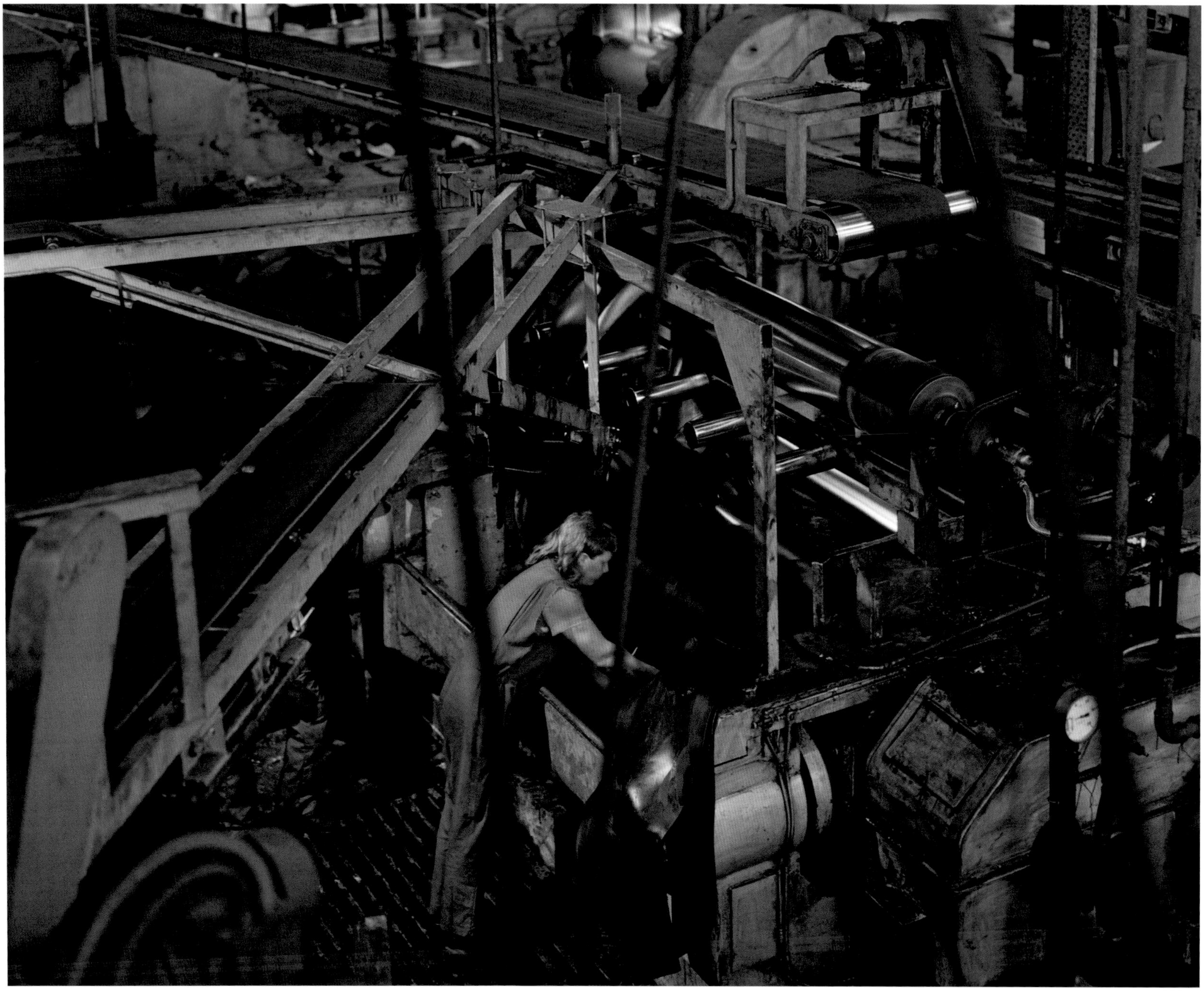

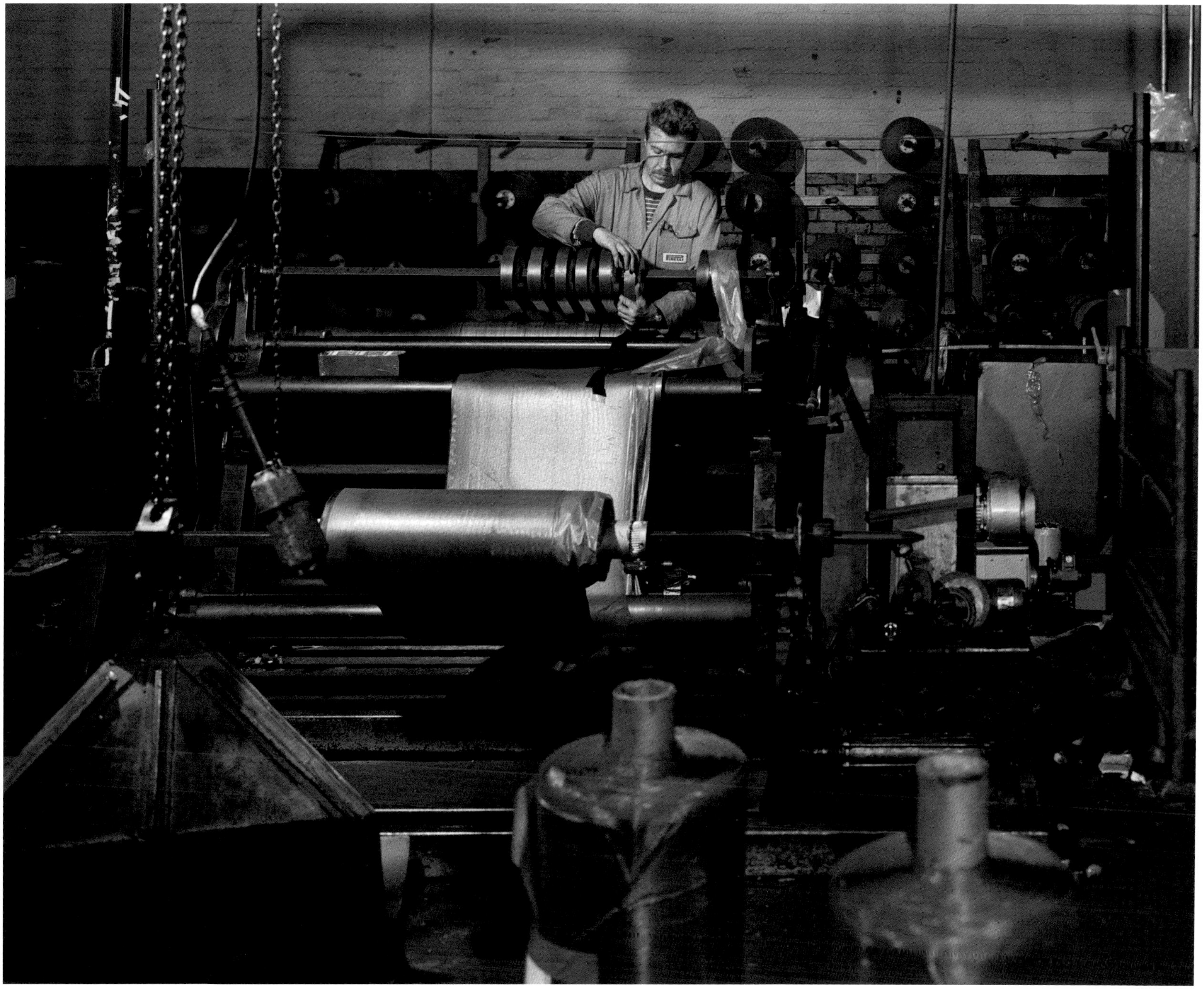

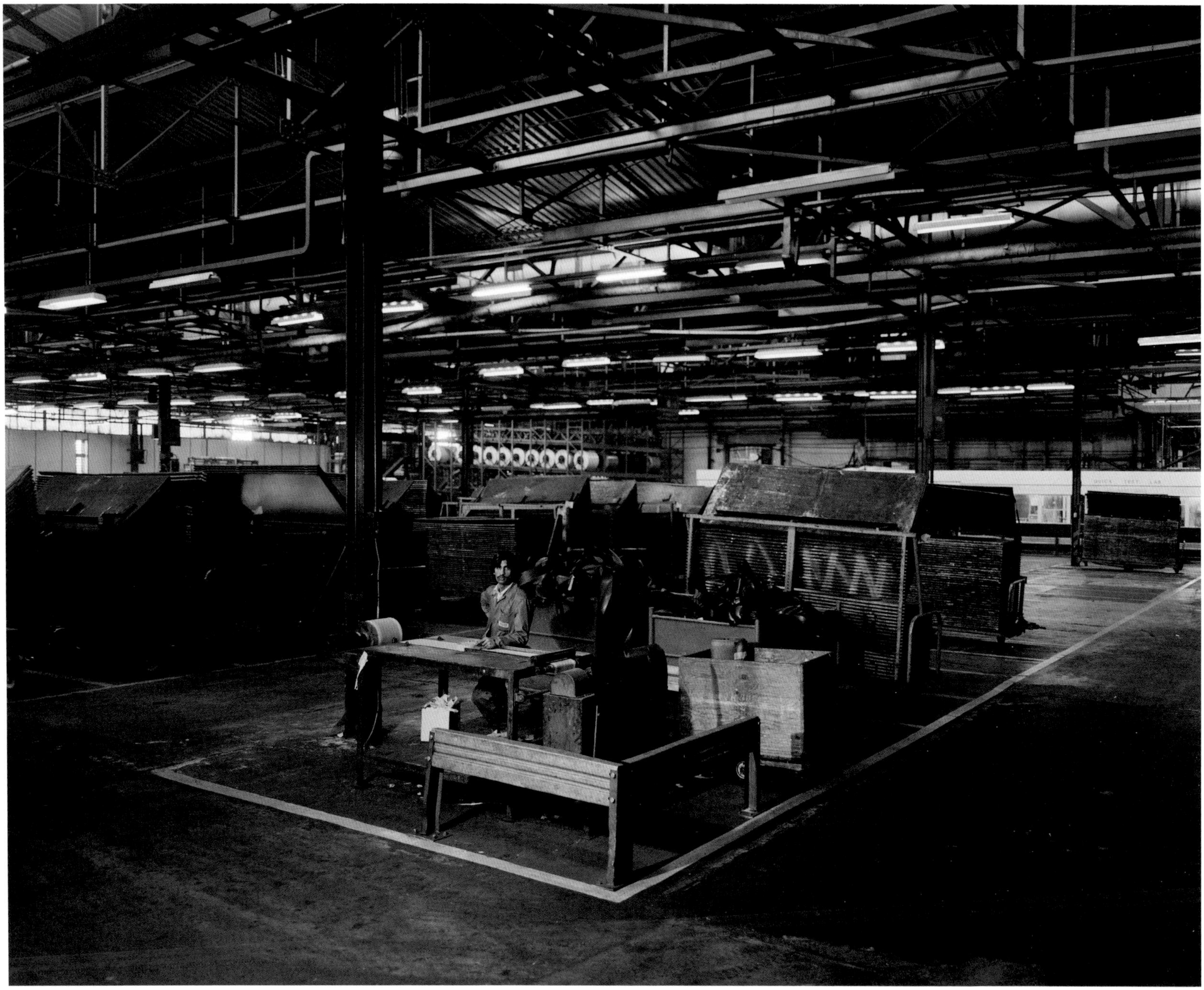

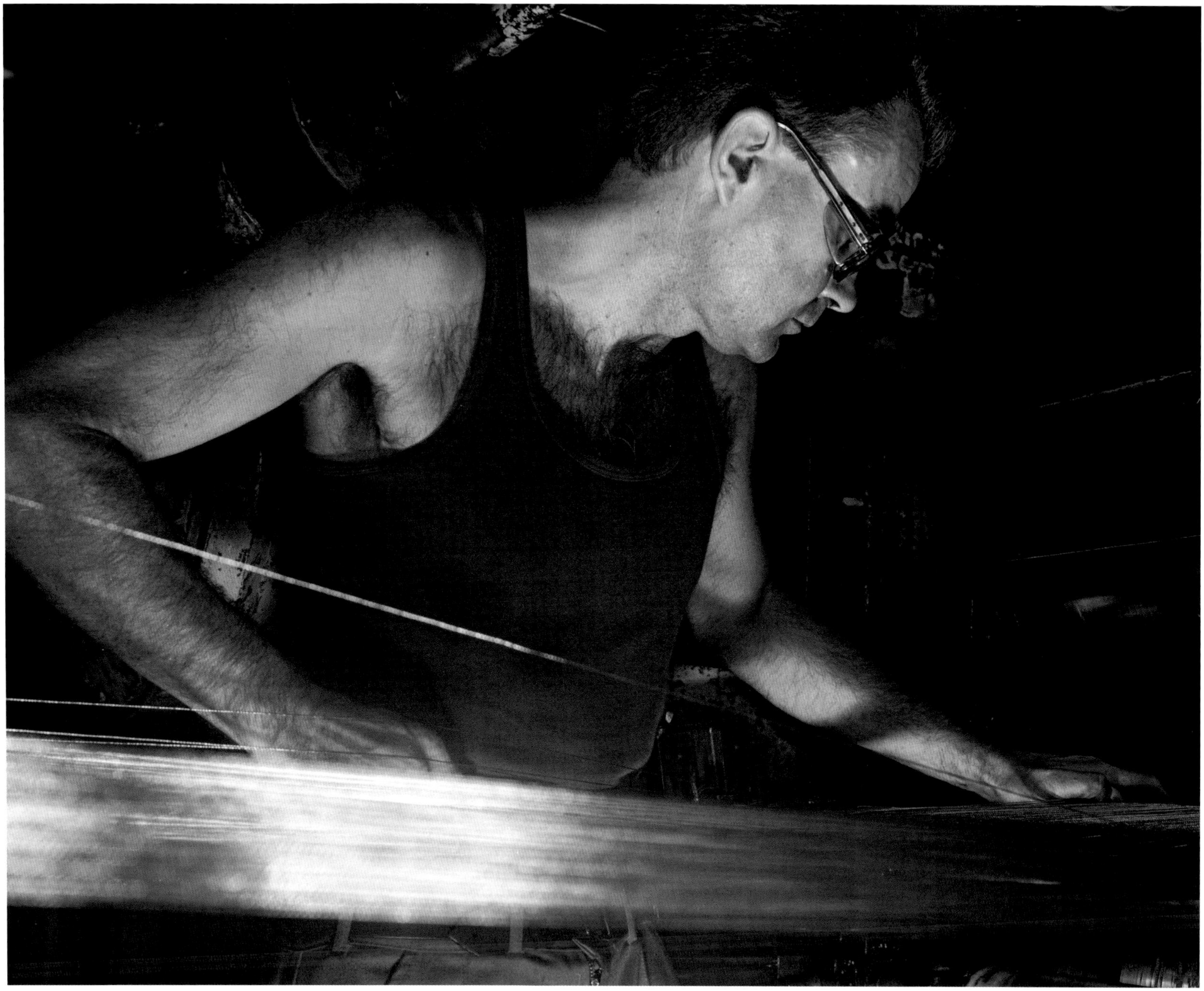

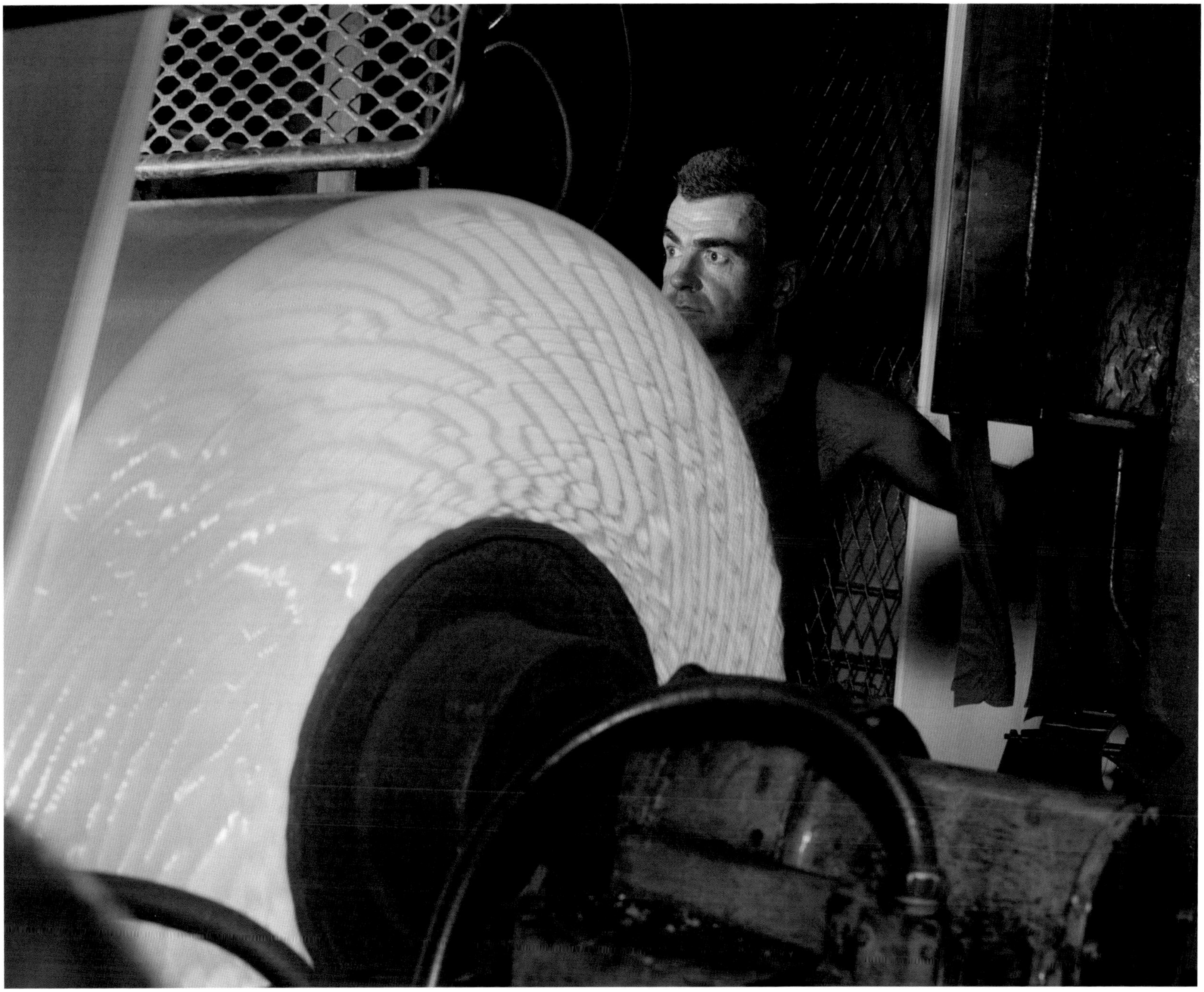

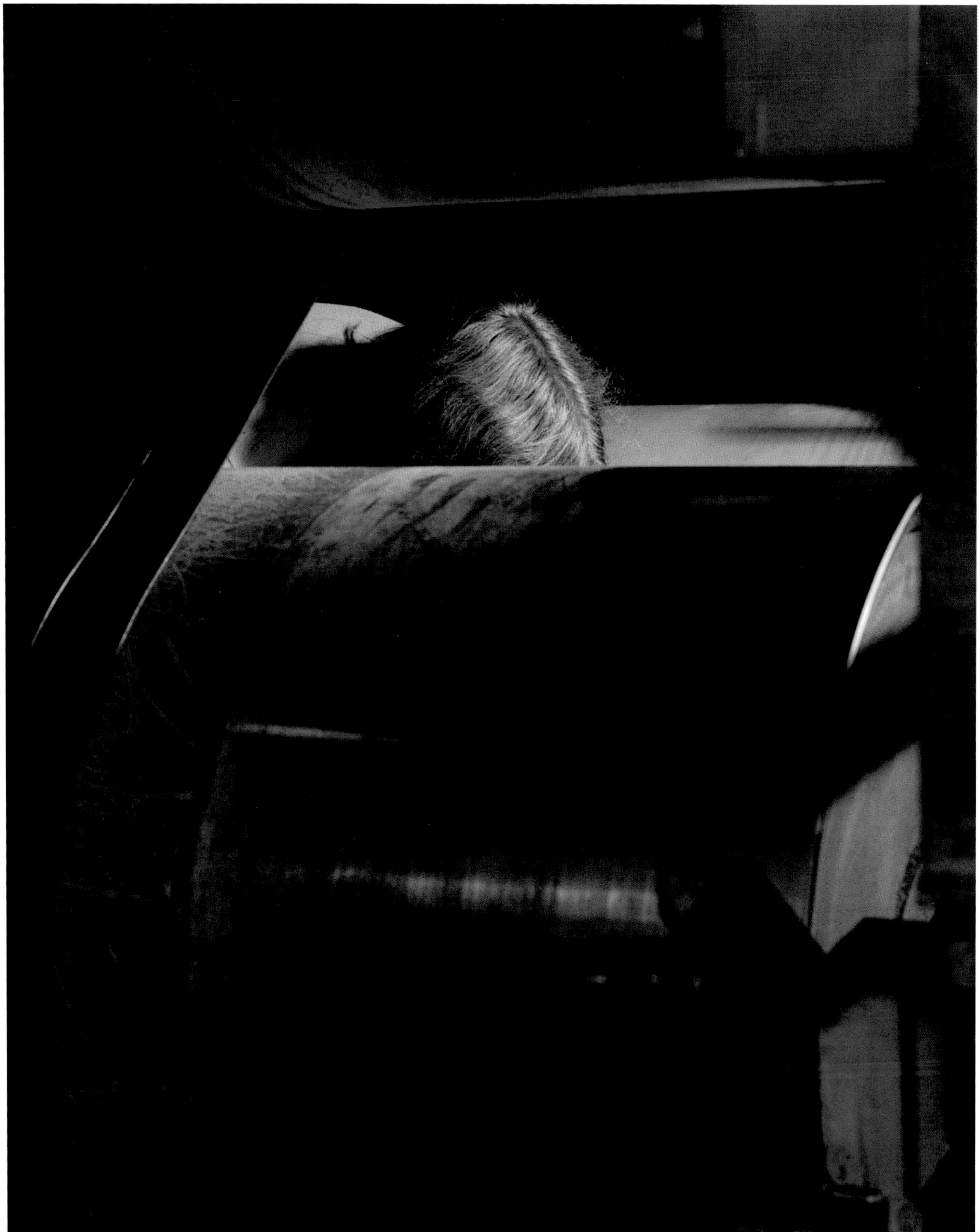

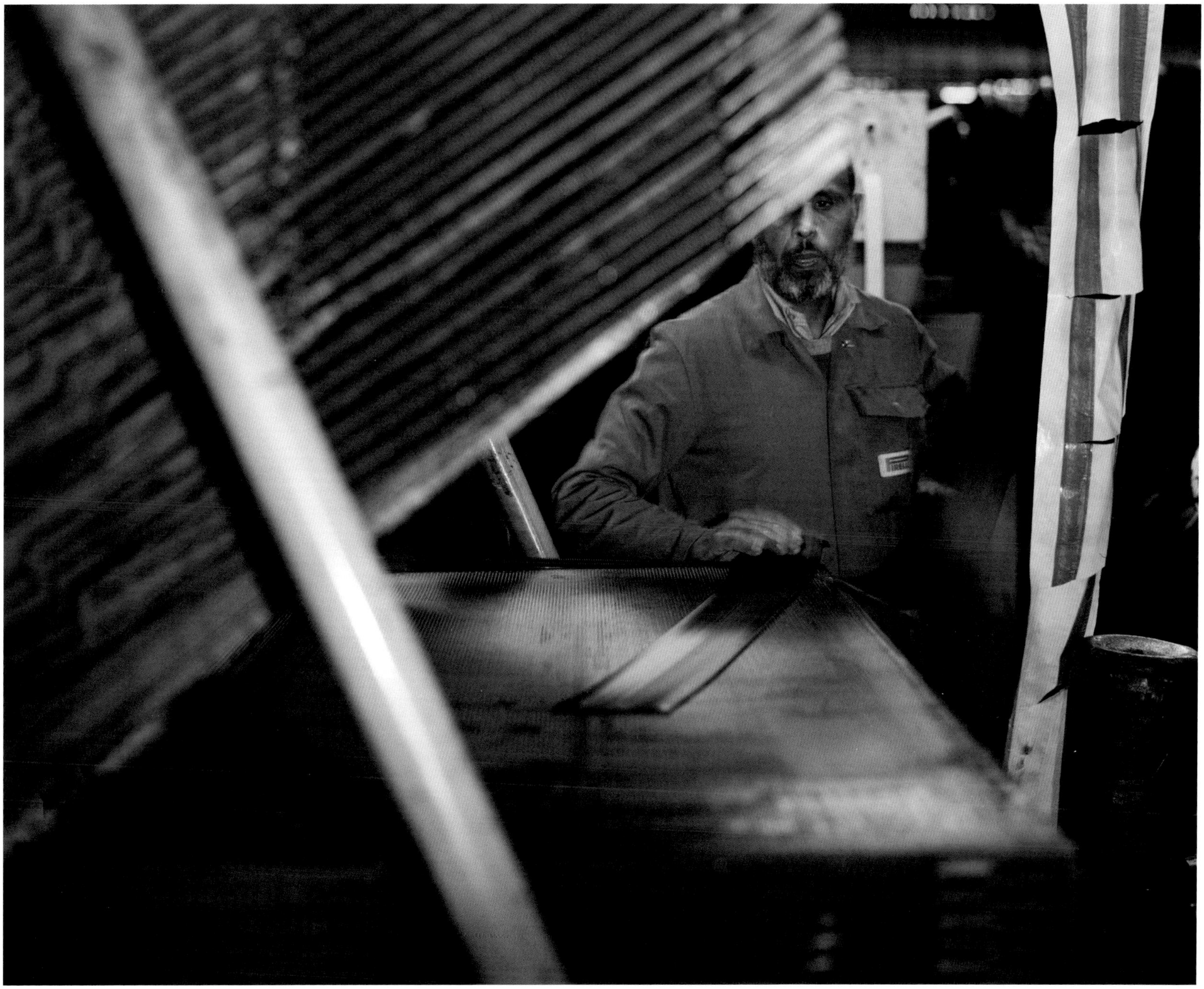

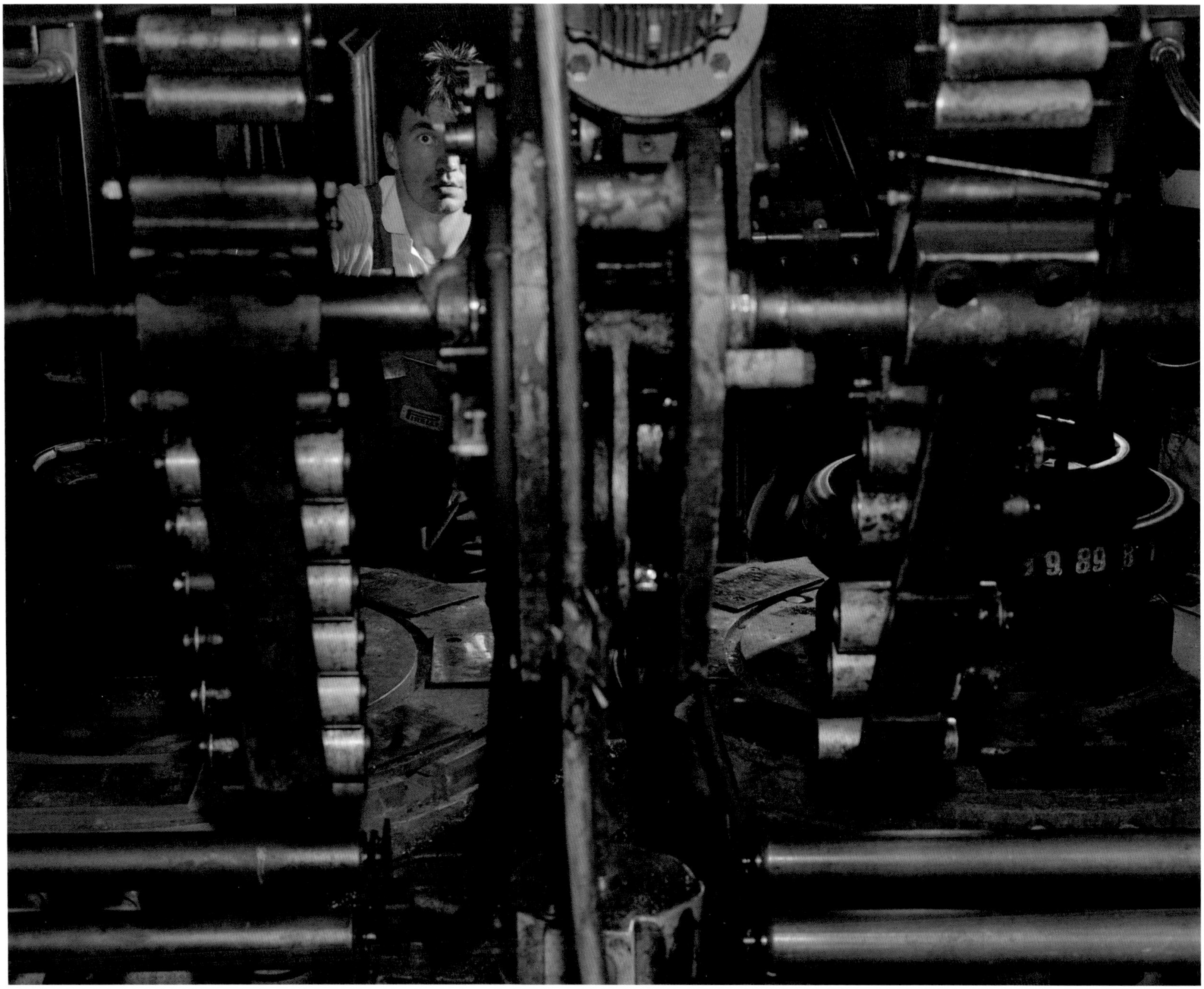

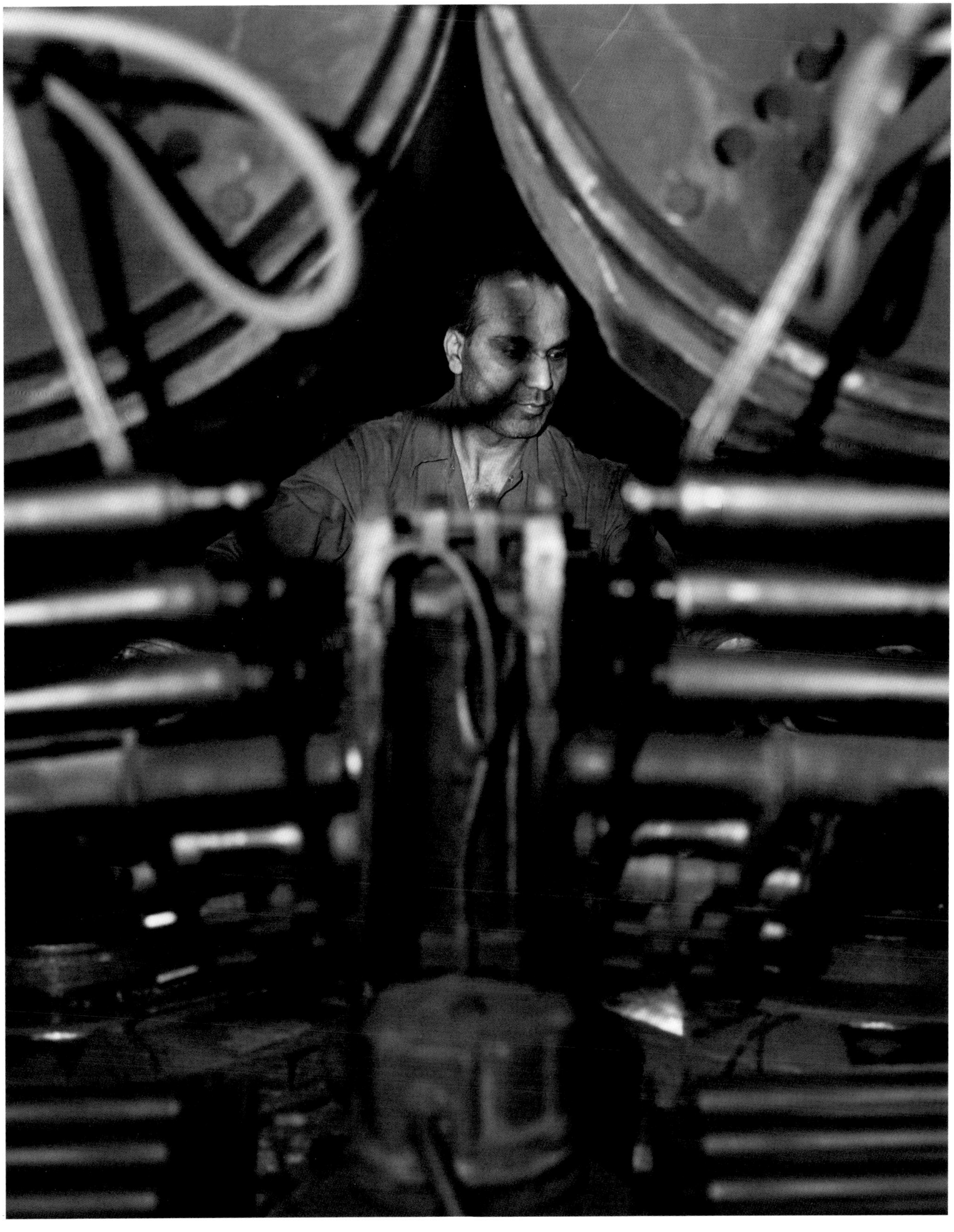

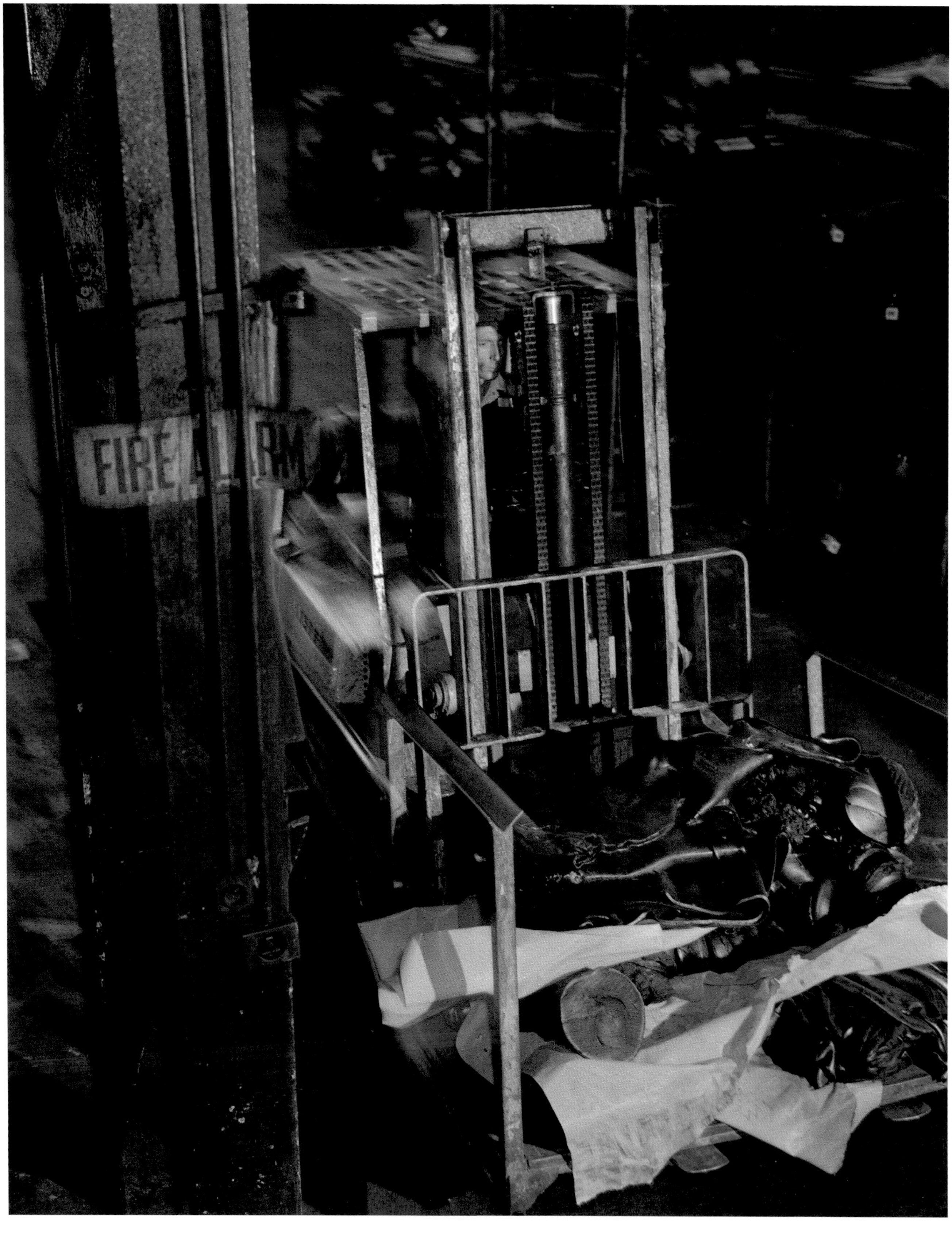
FIRE

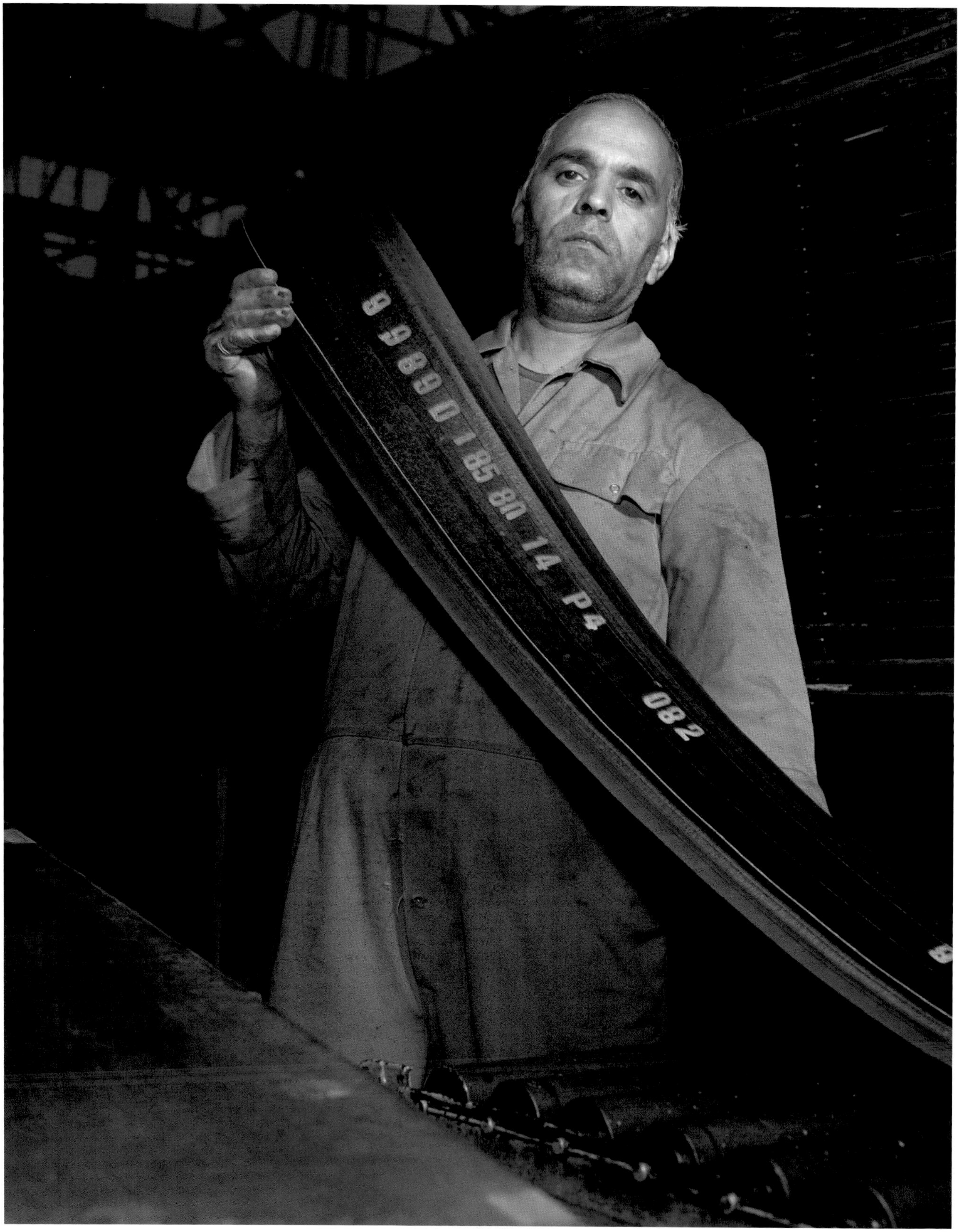

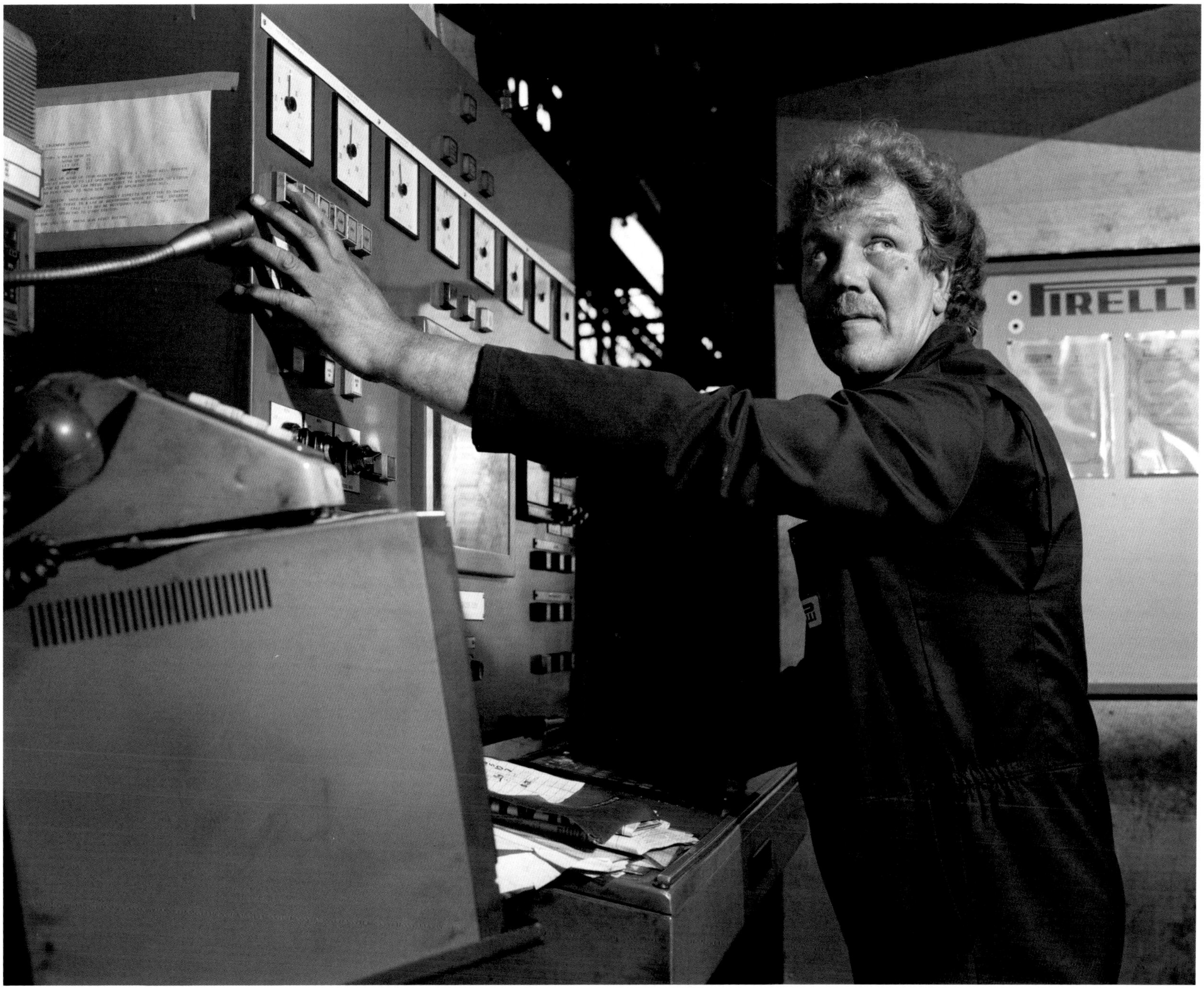
PIRELLI

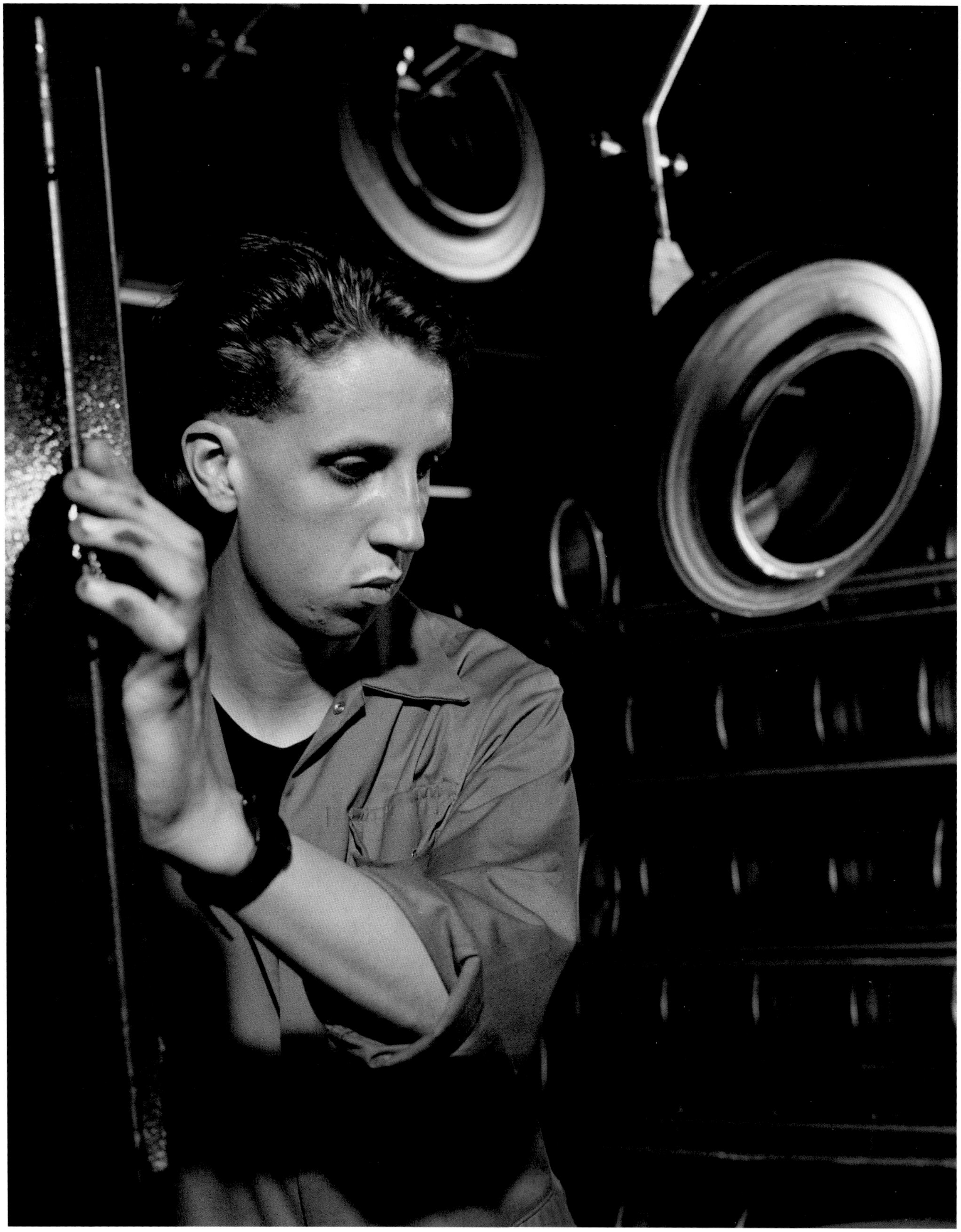

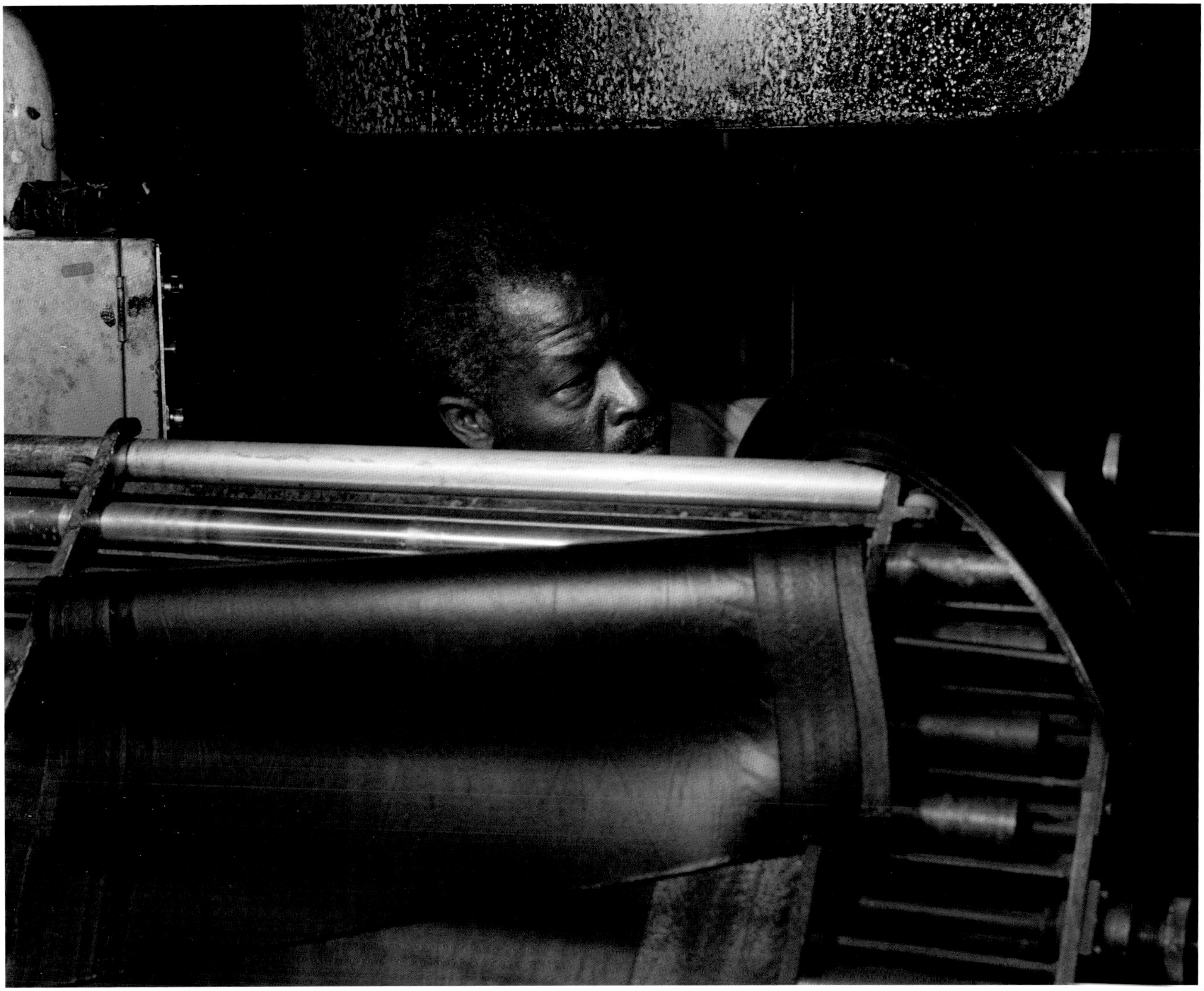

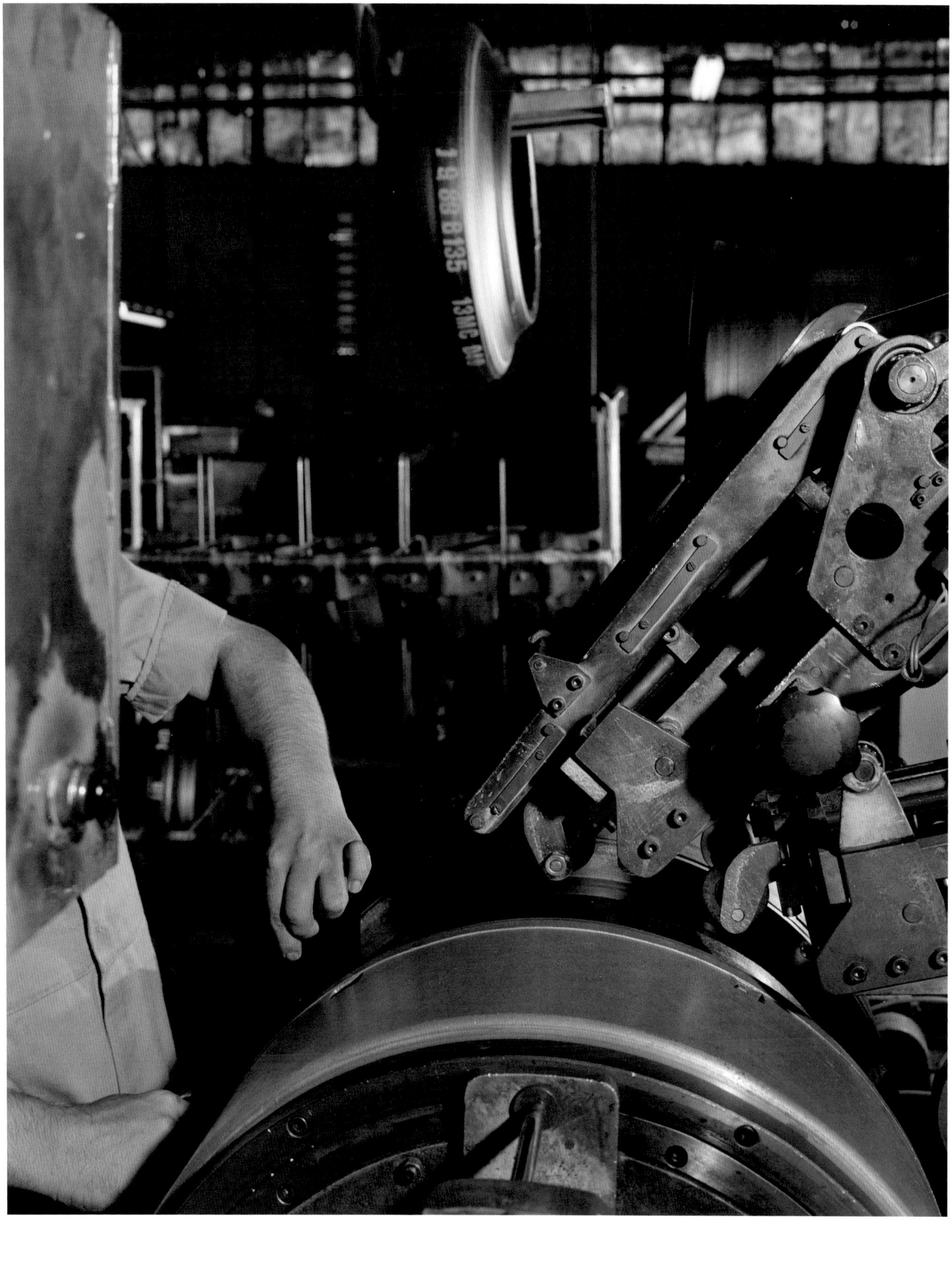

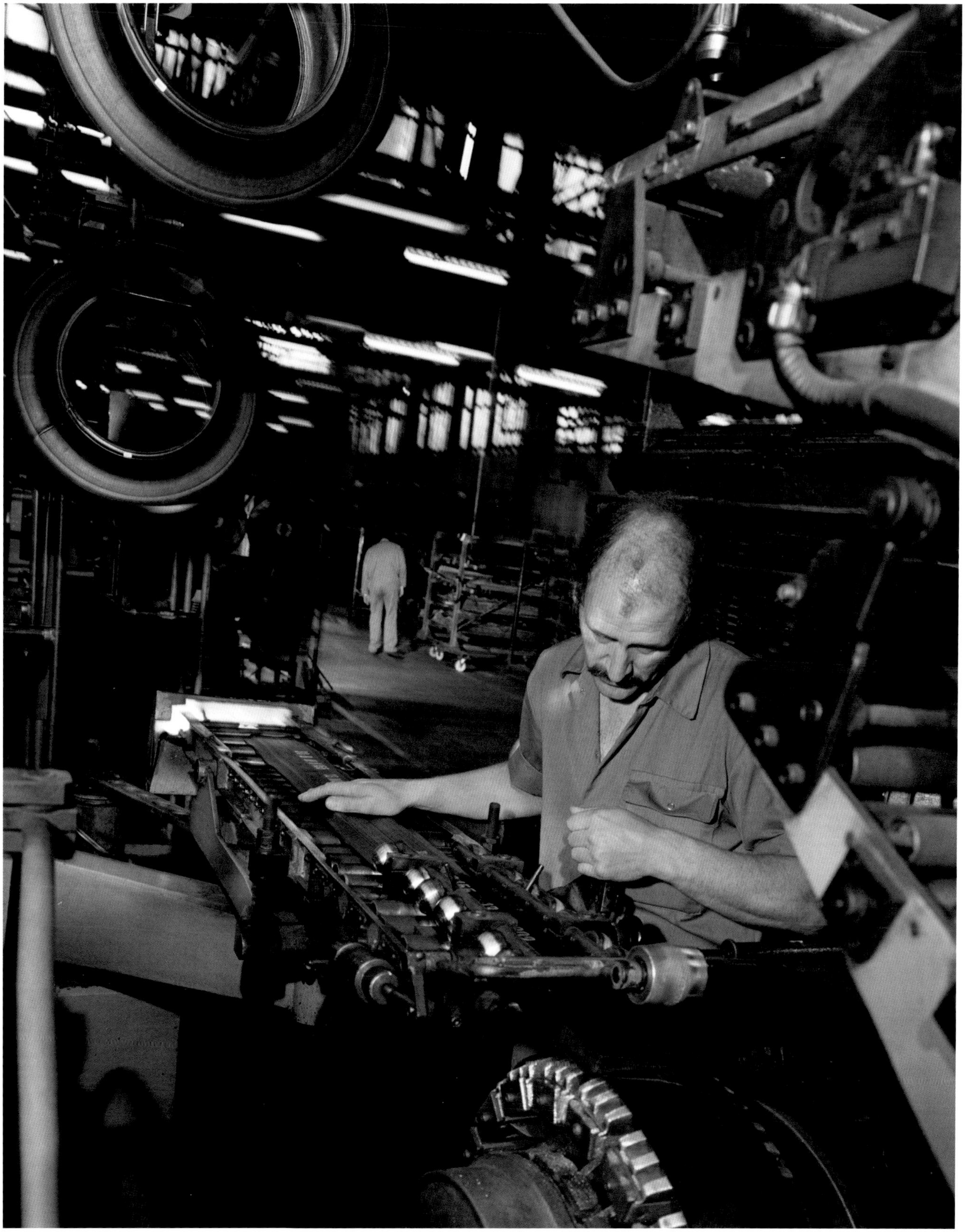

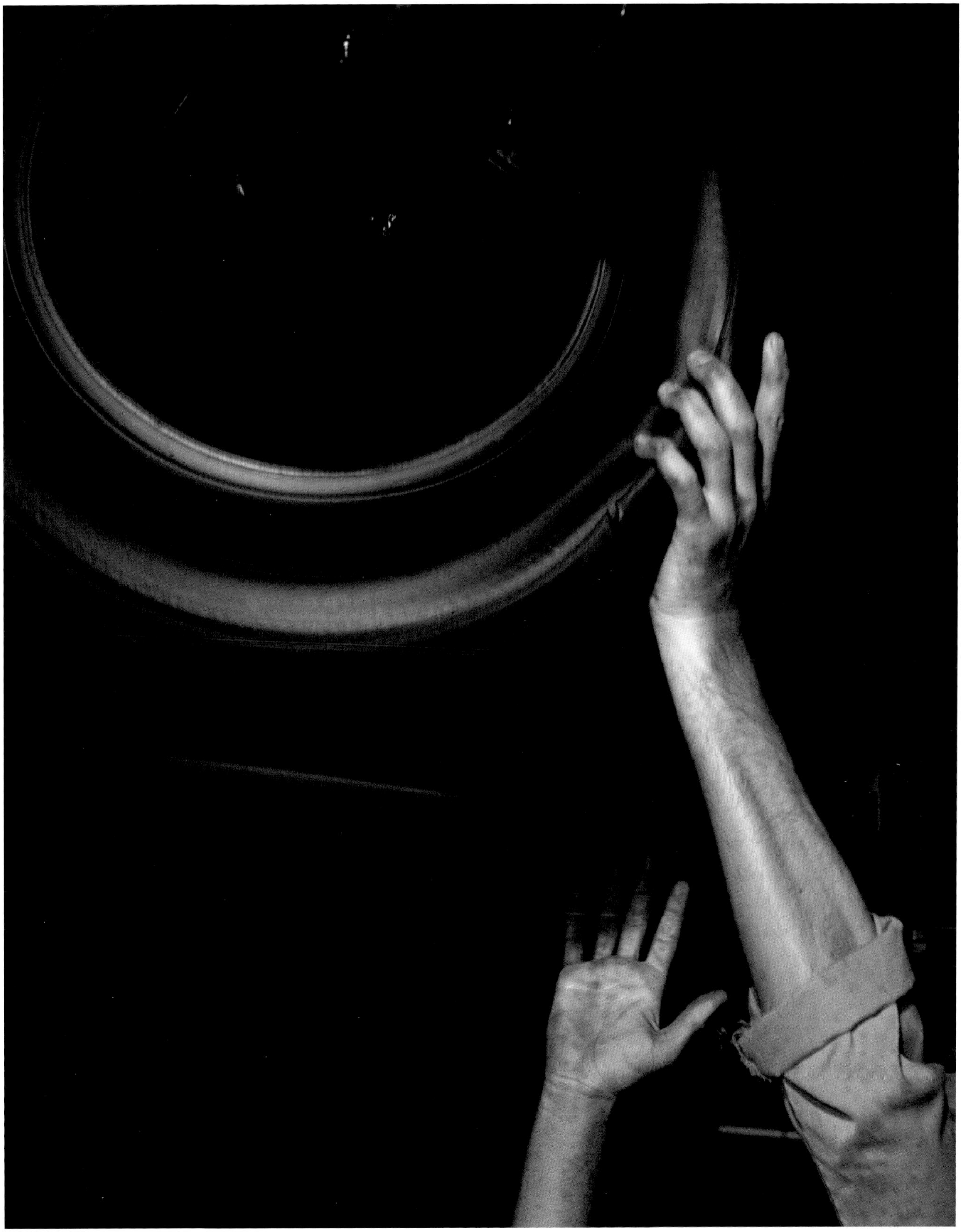

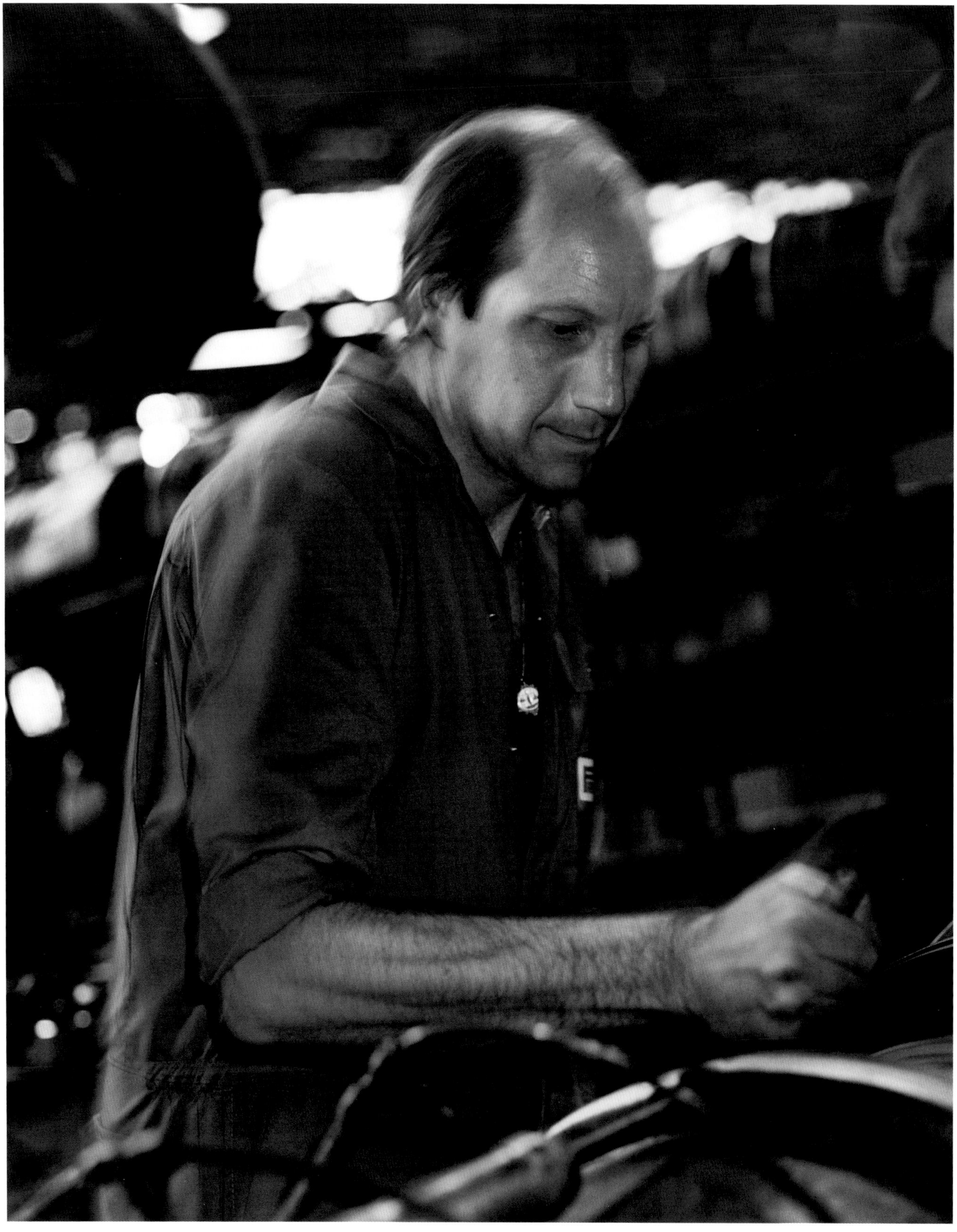

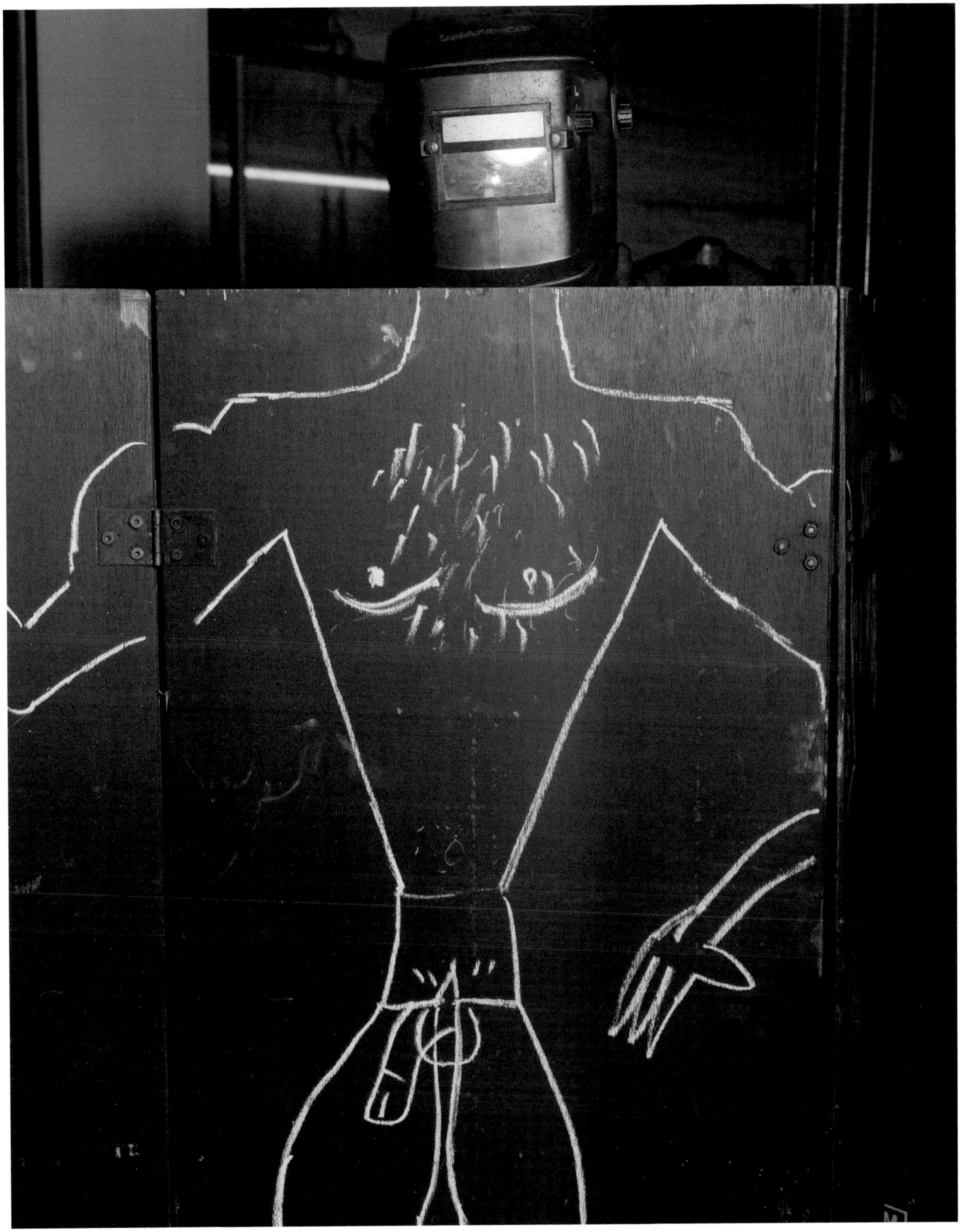

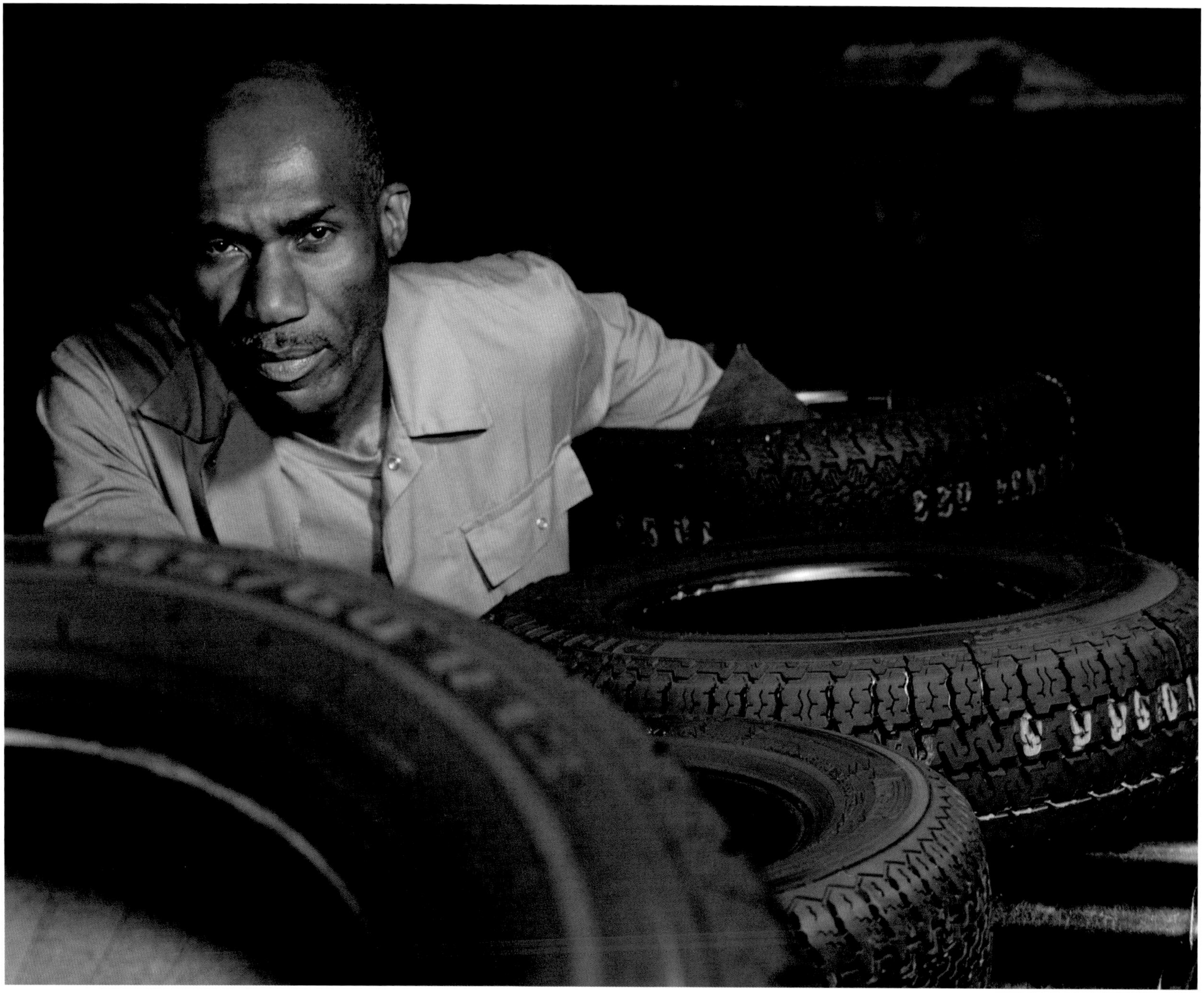

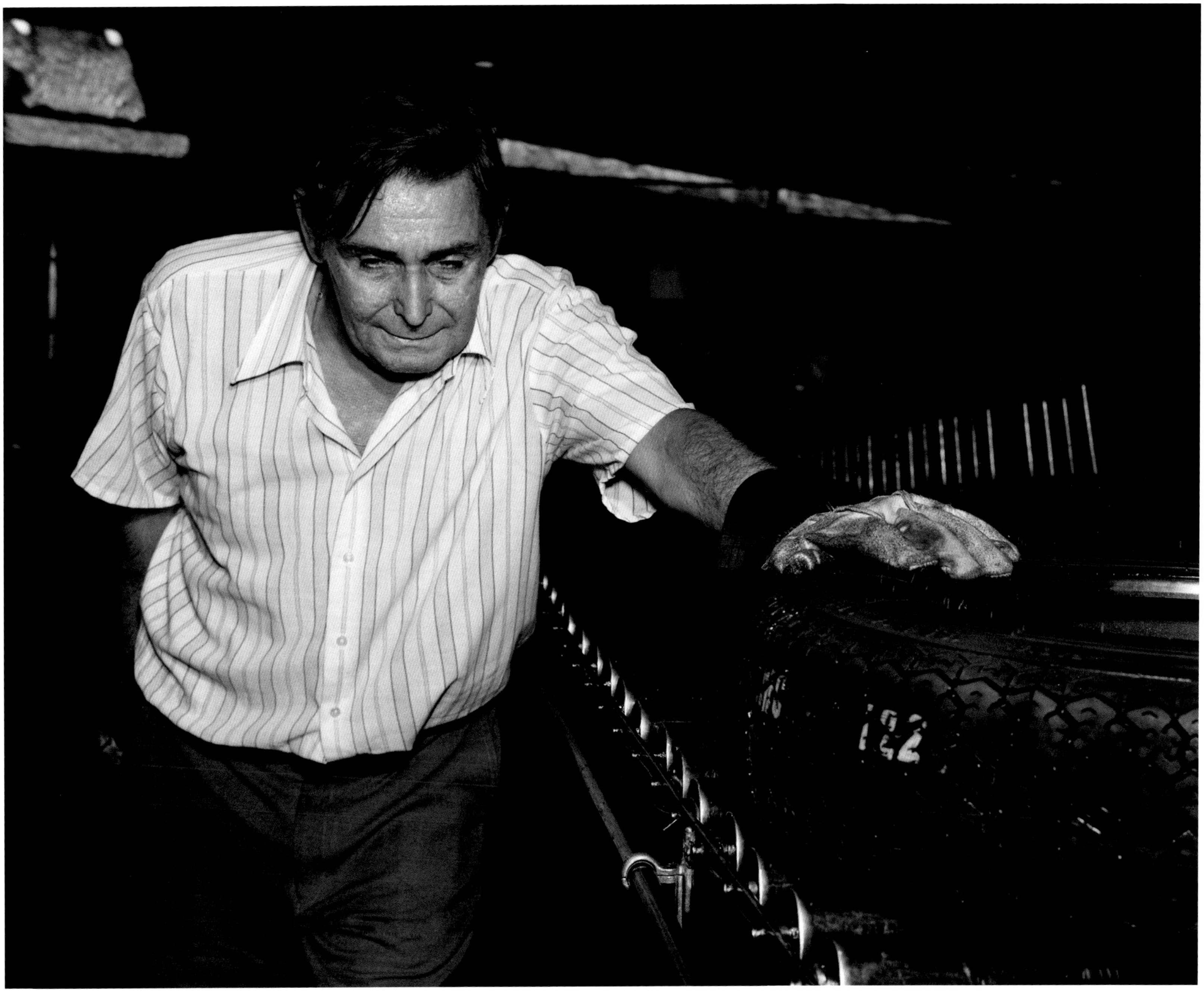

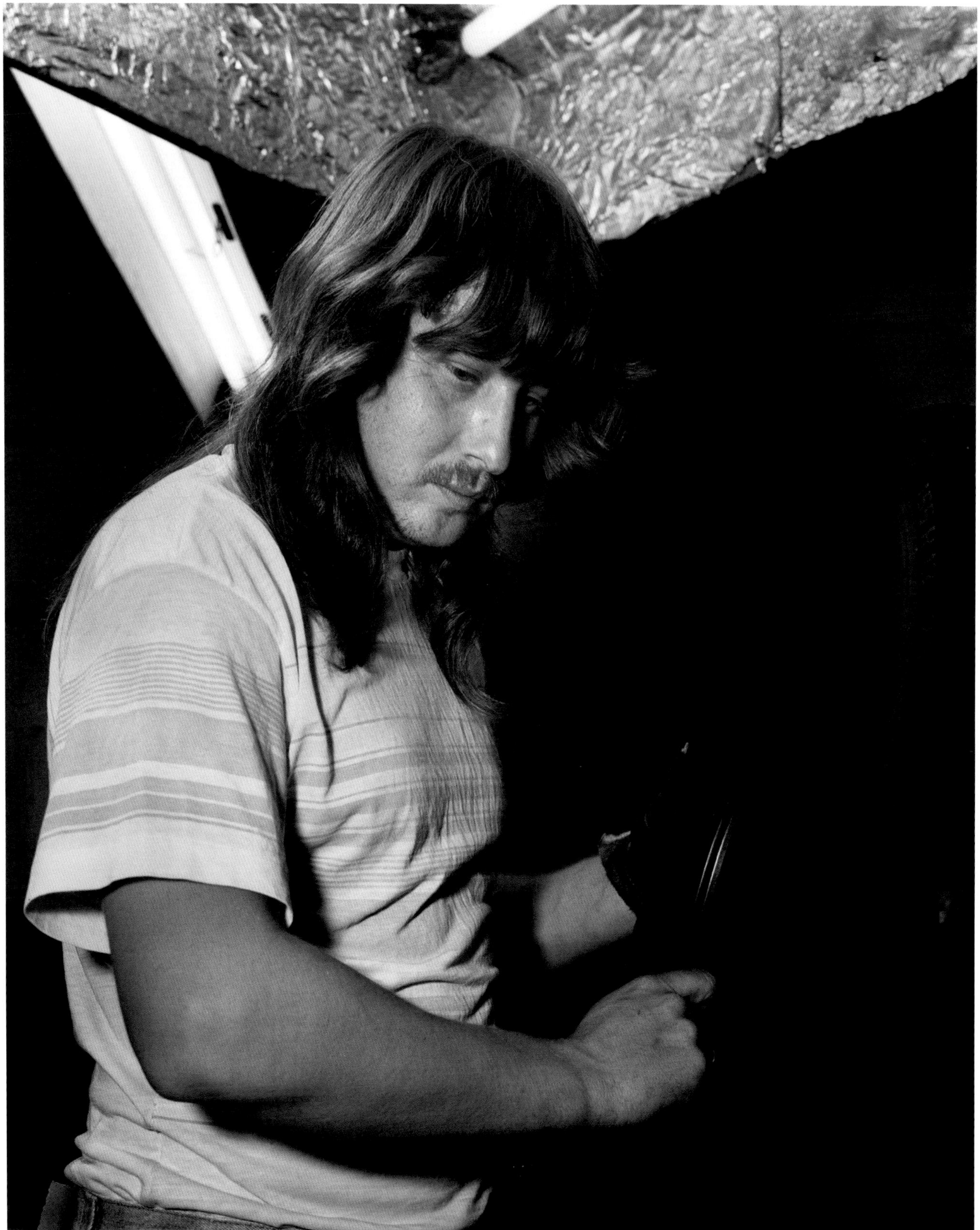

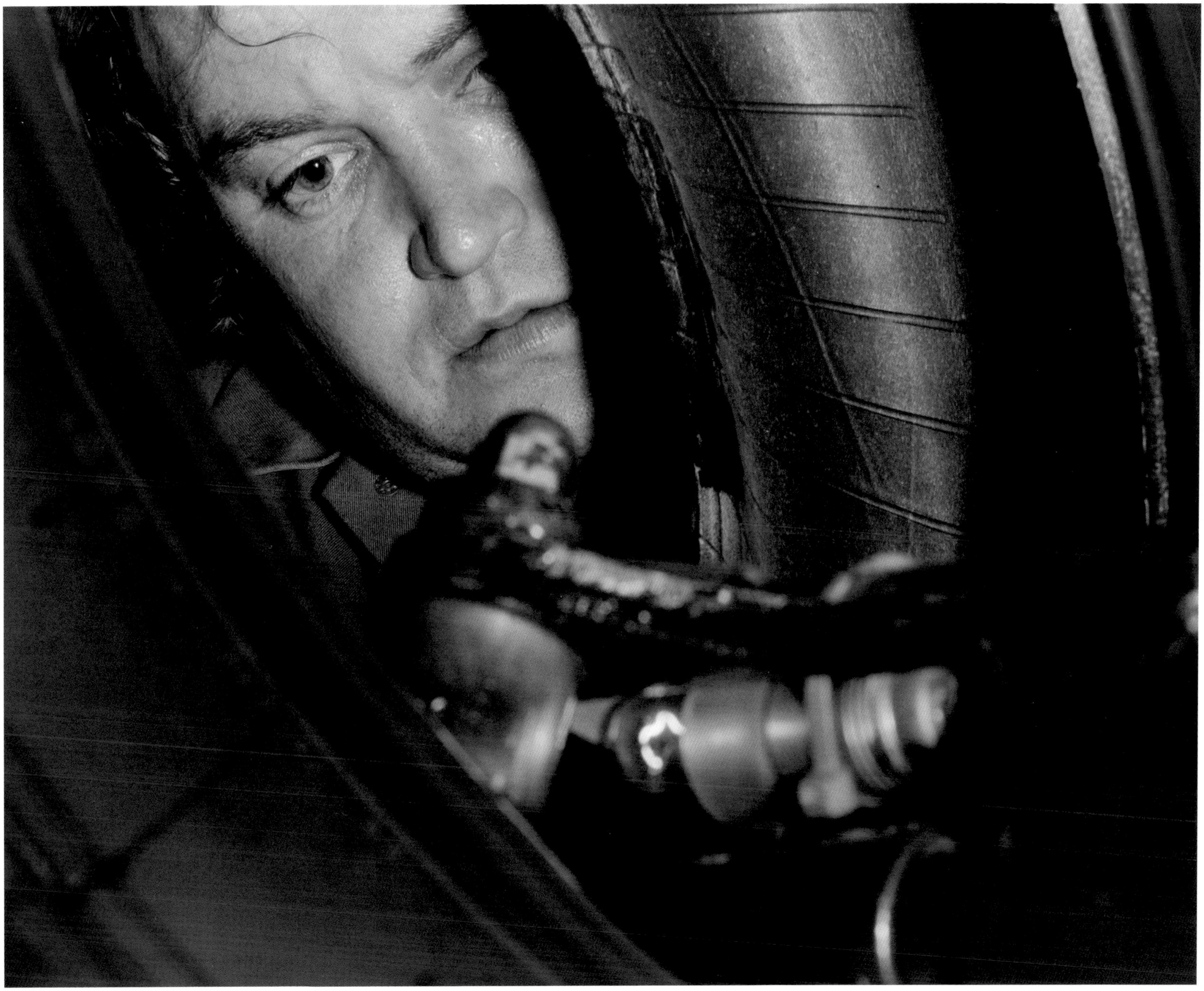

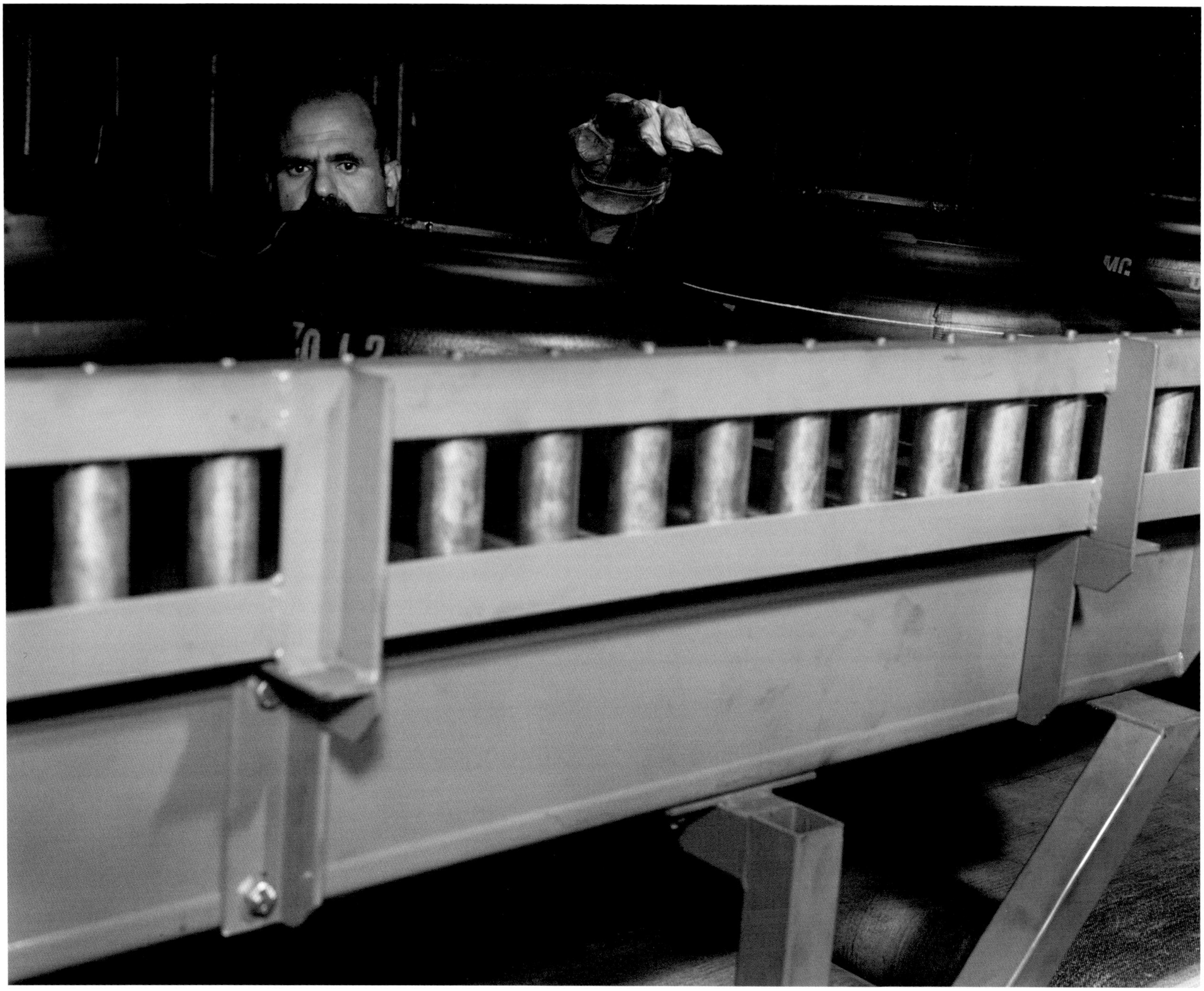

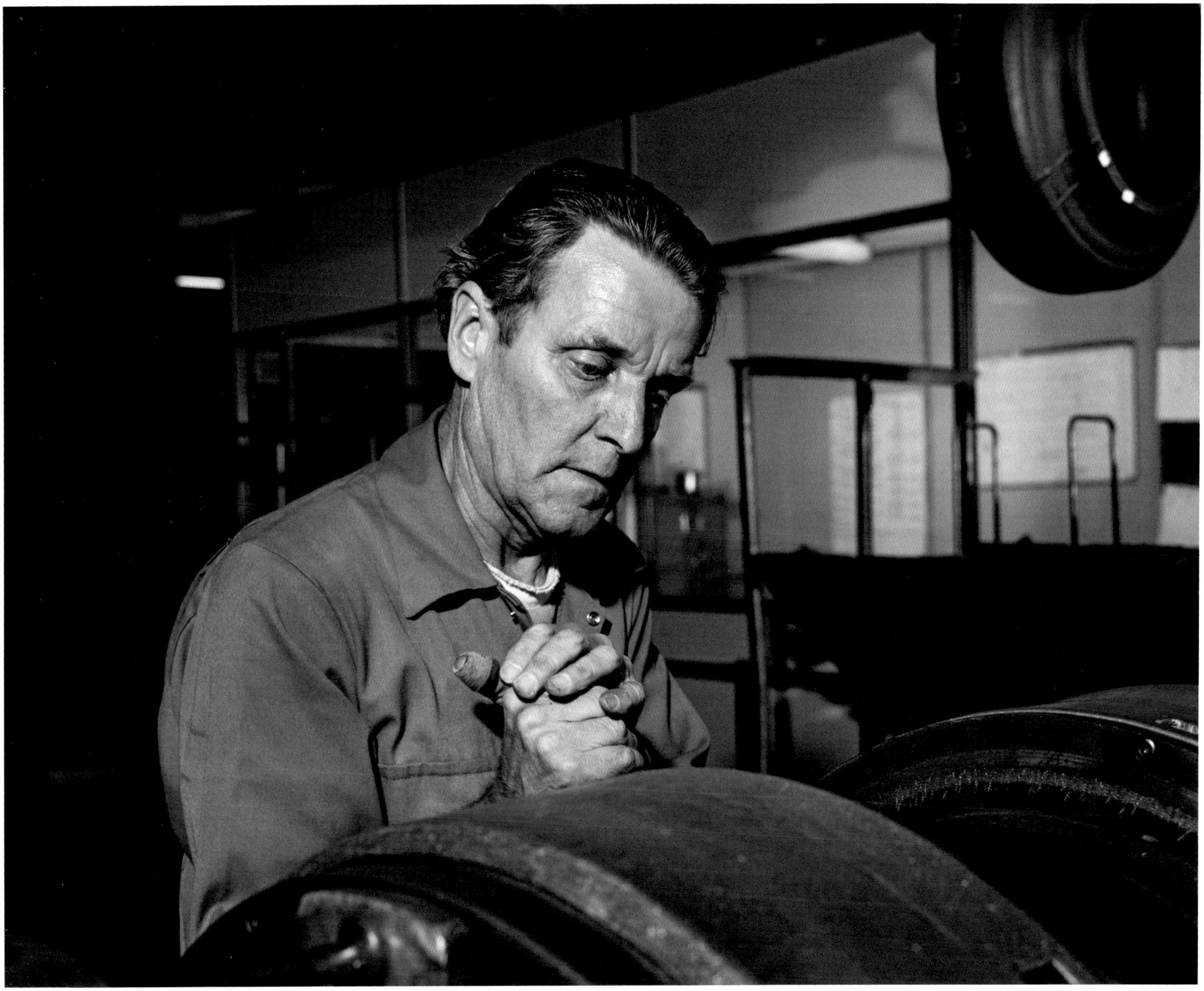

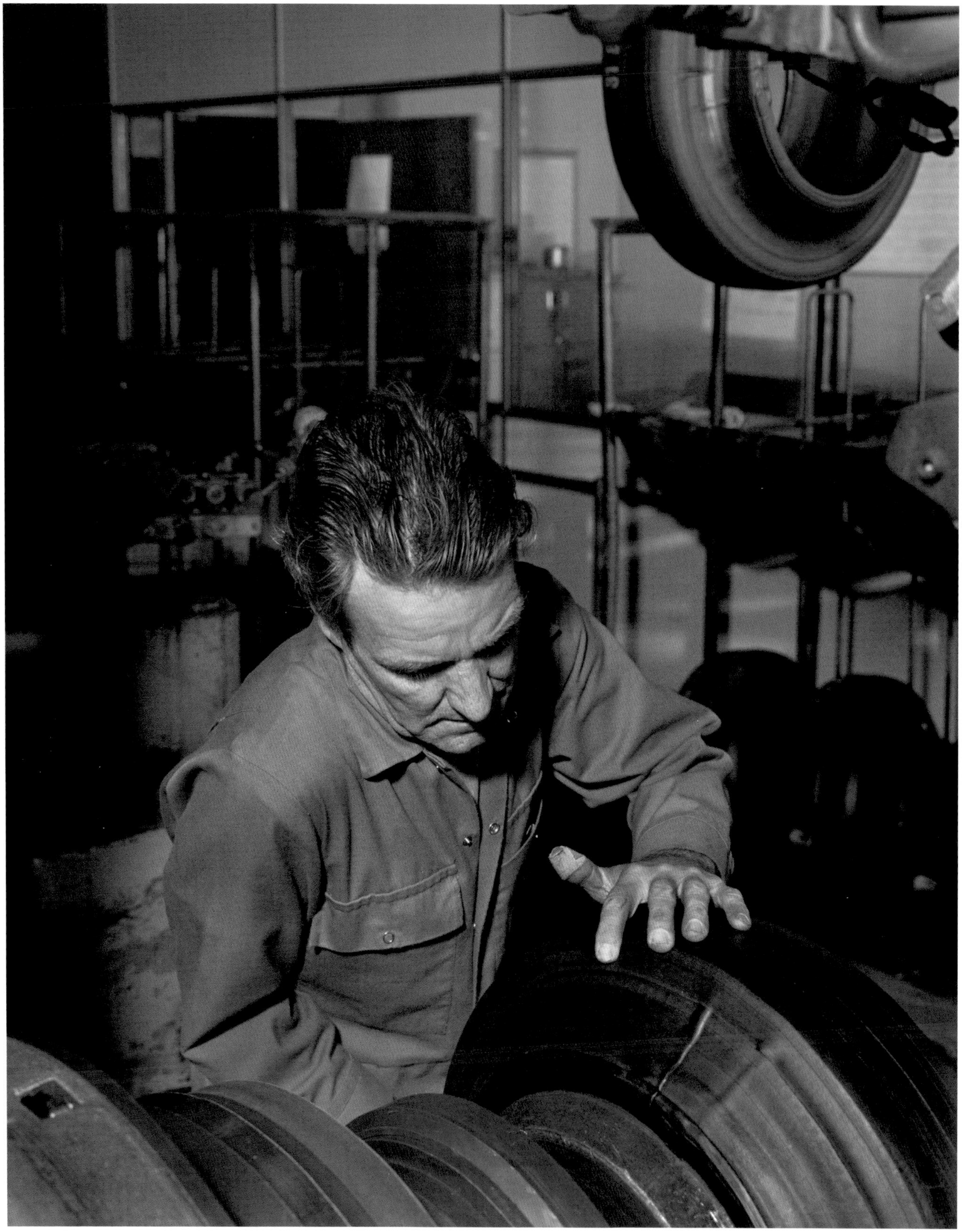

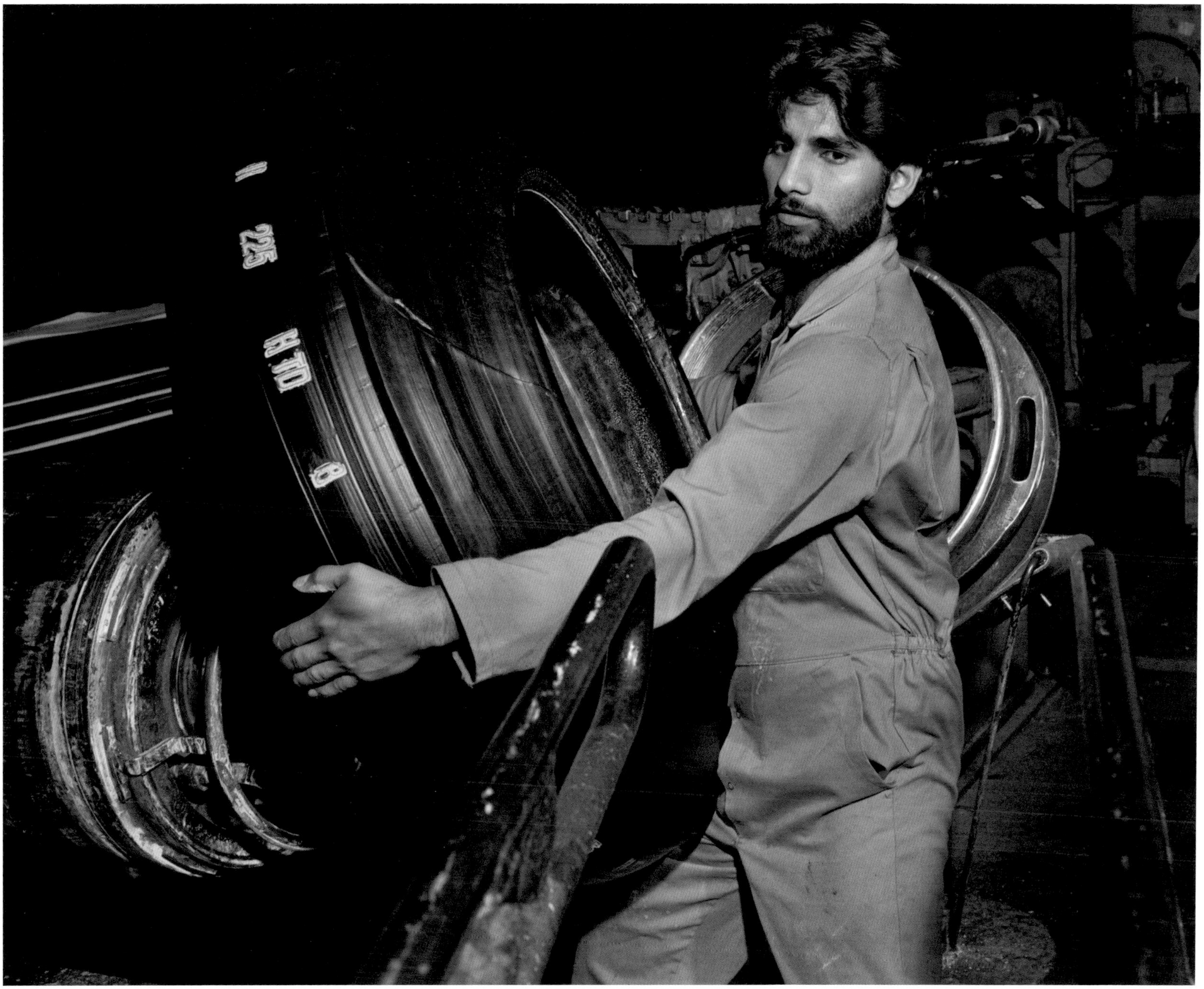

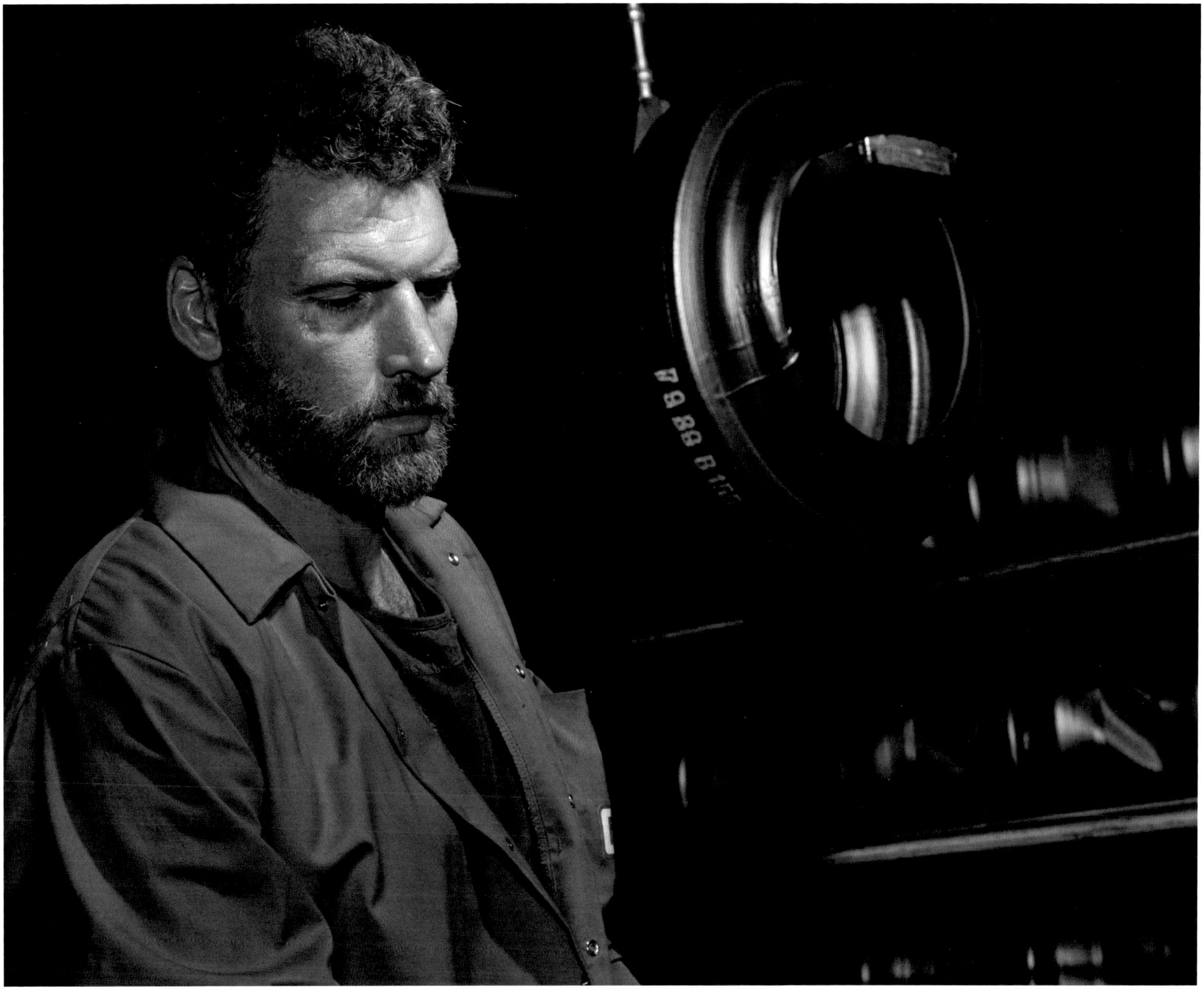

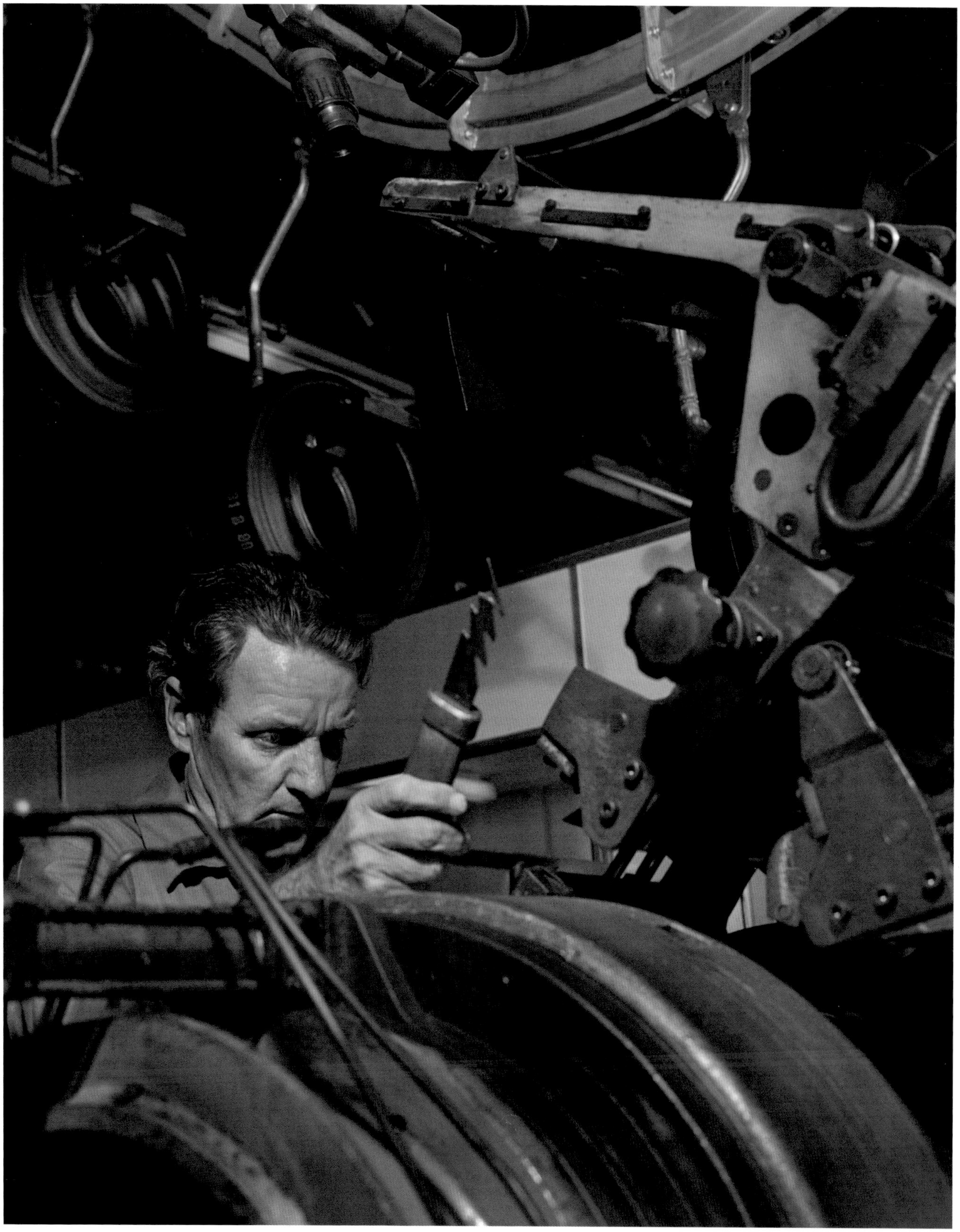

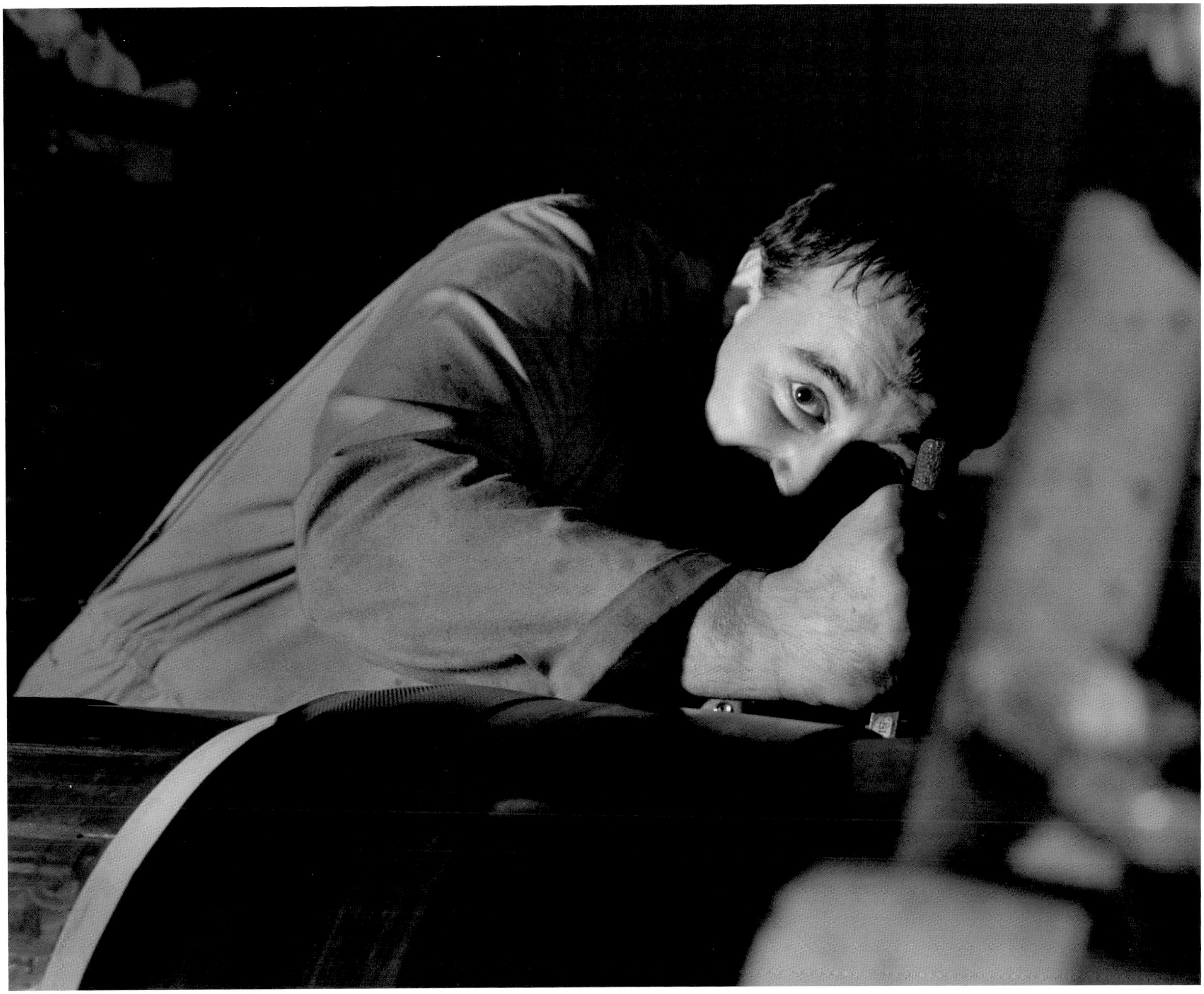

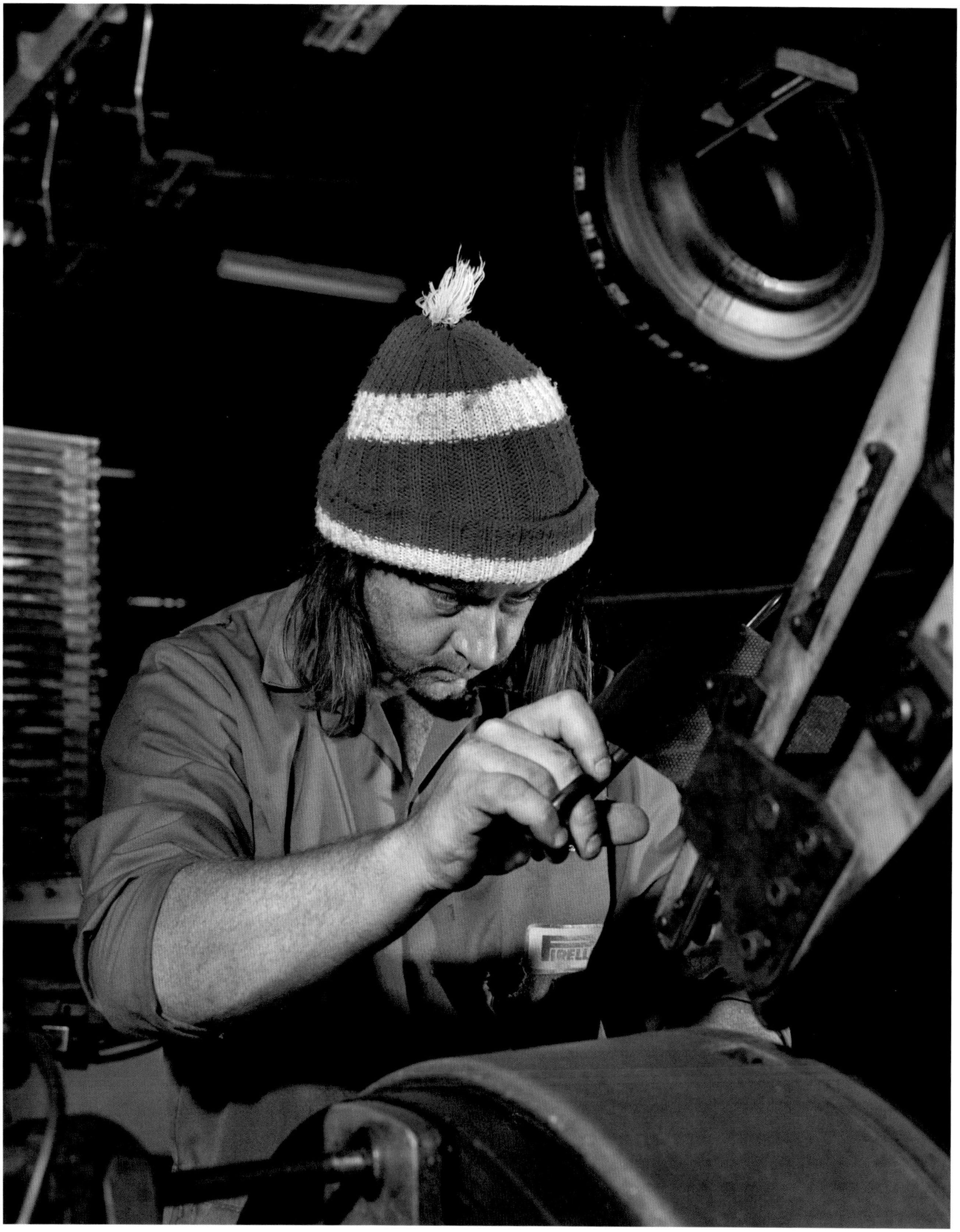

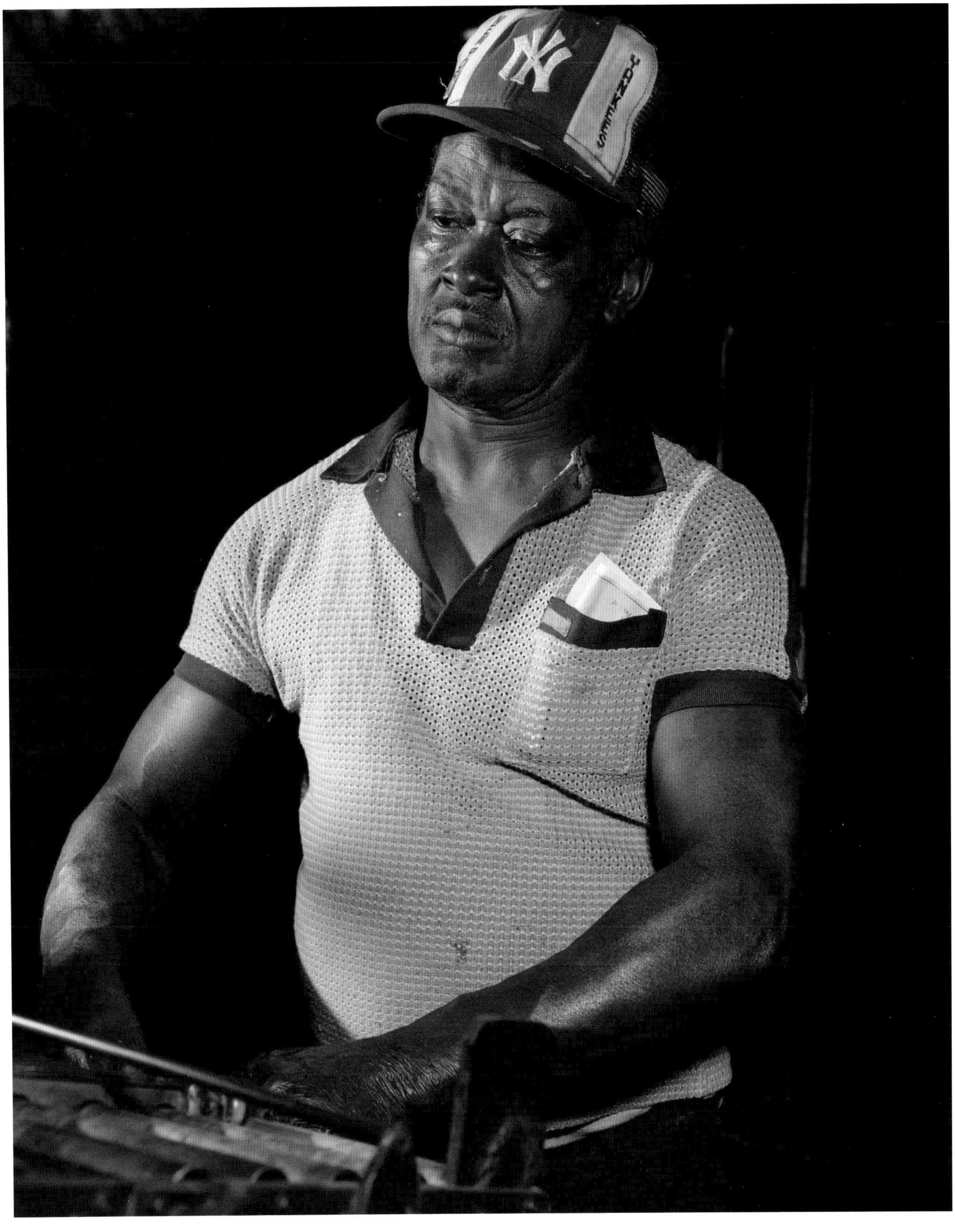
NY

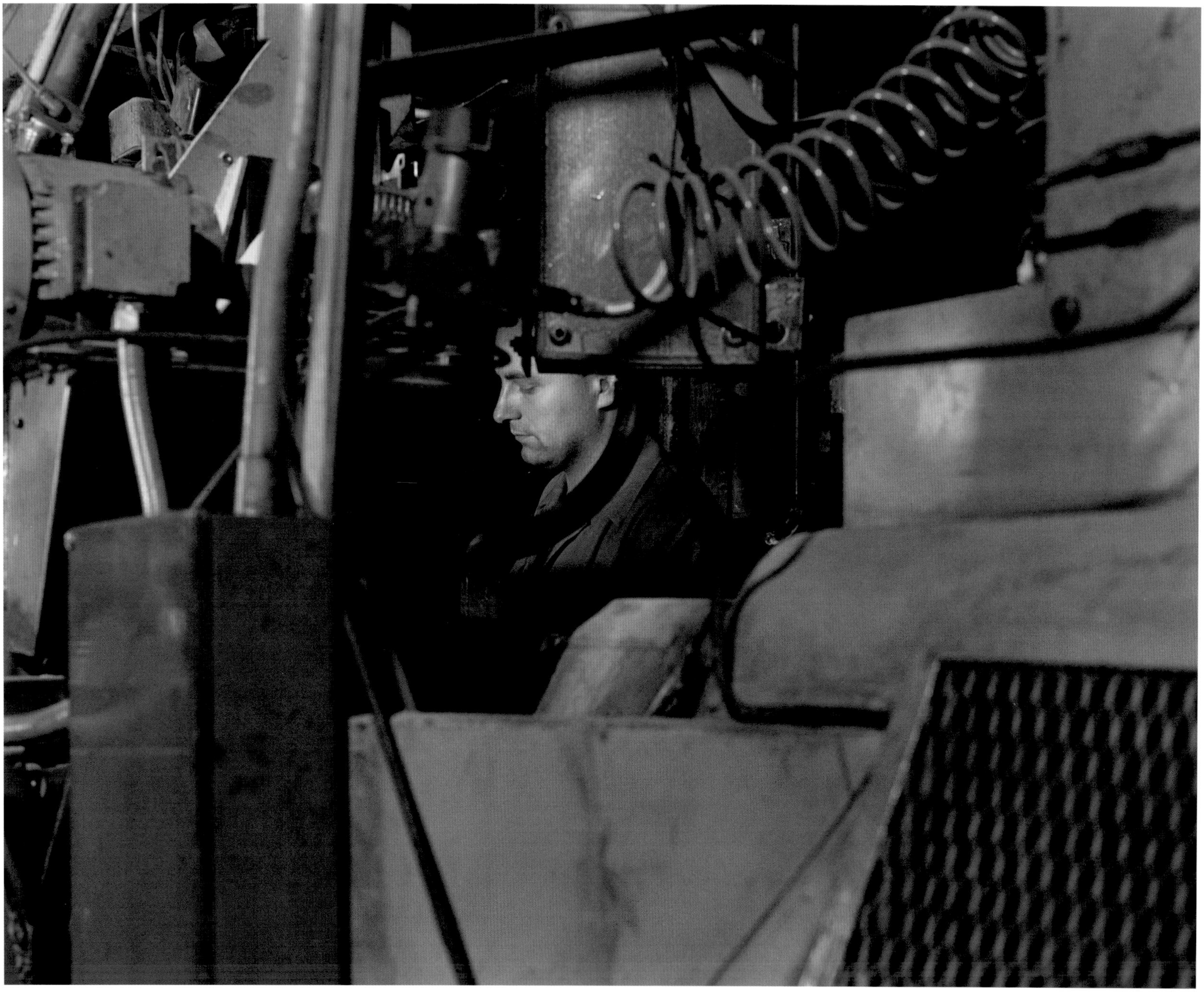

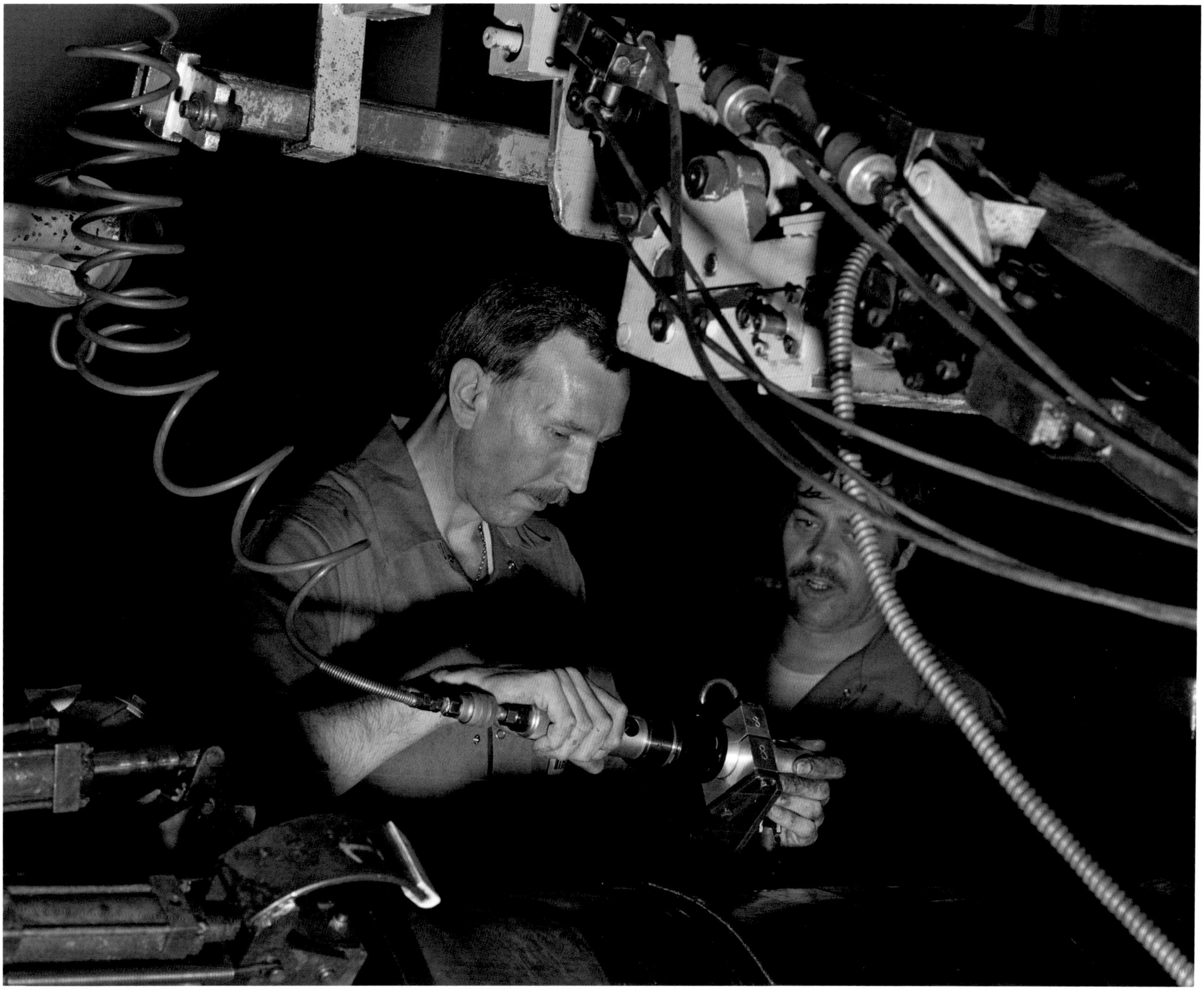

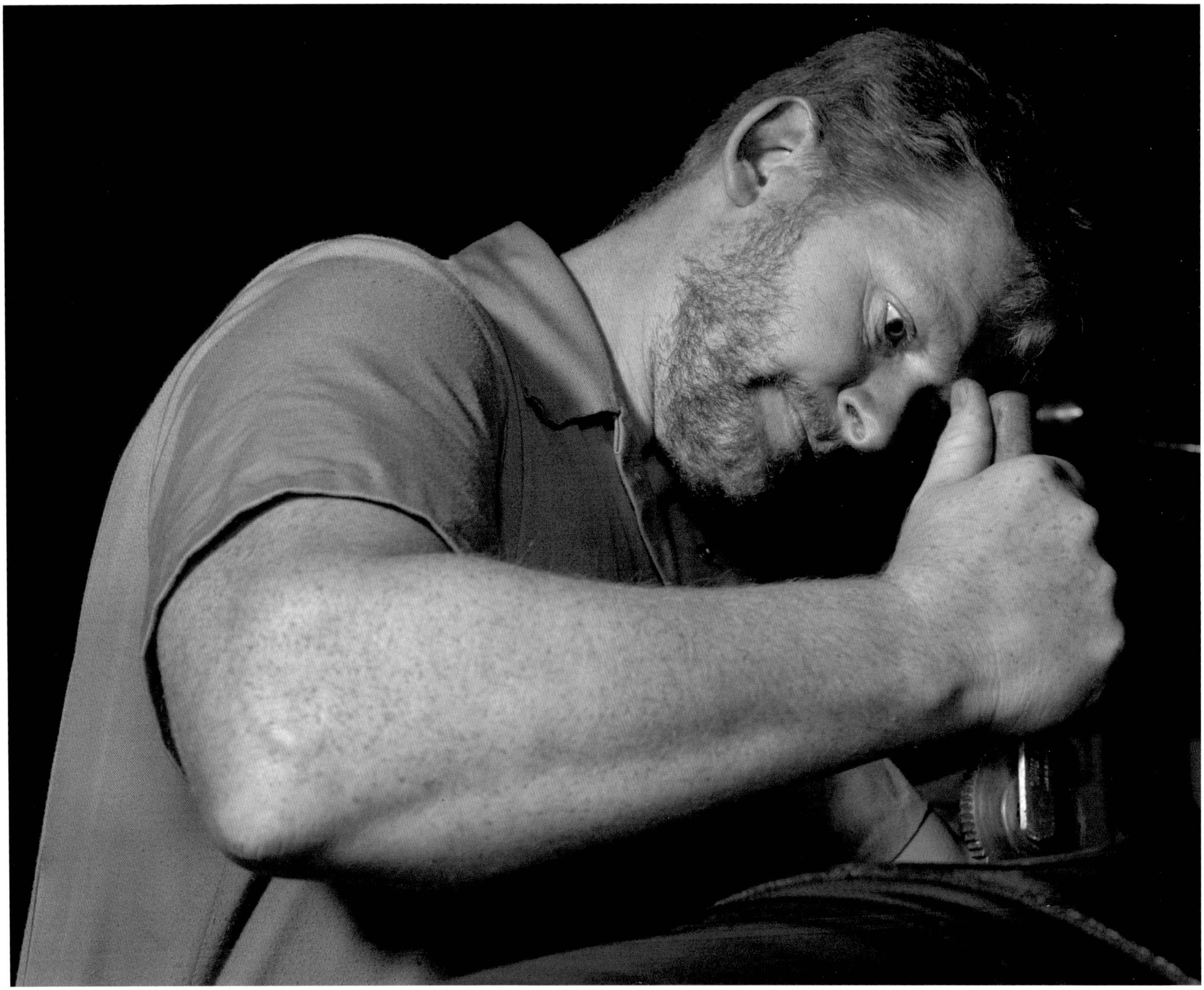

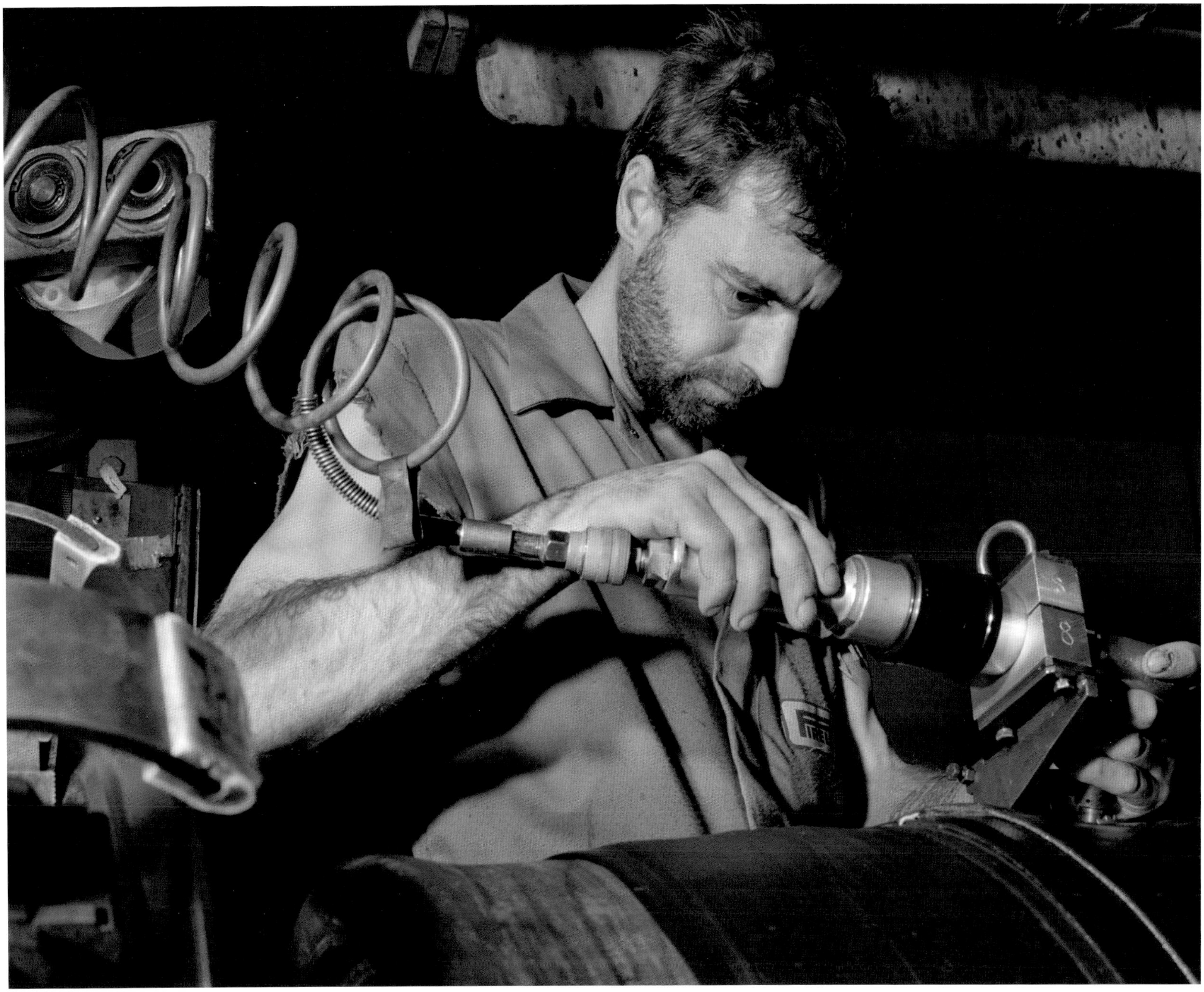

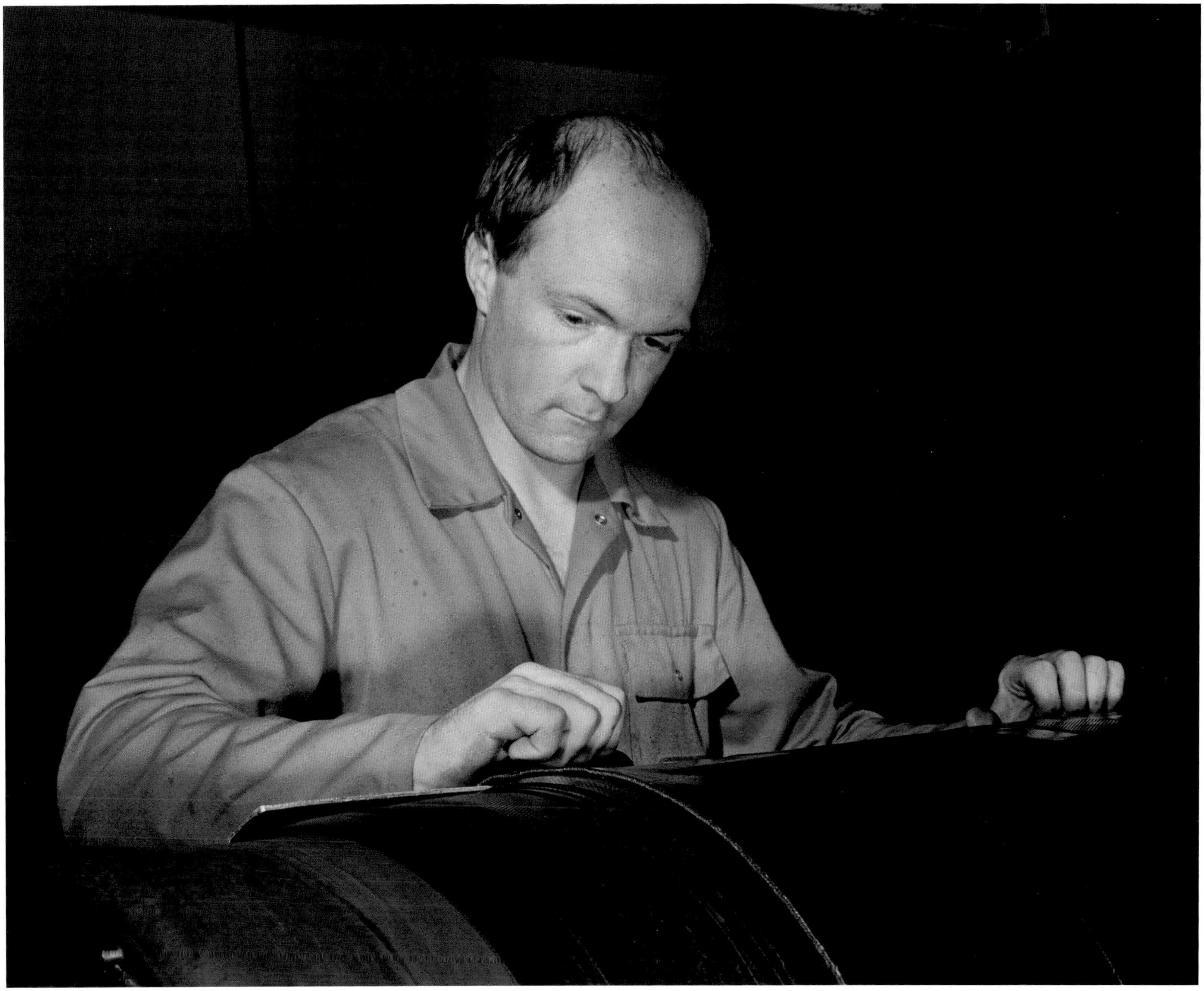

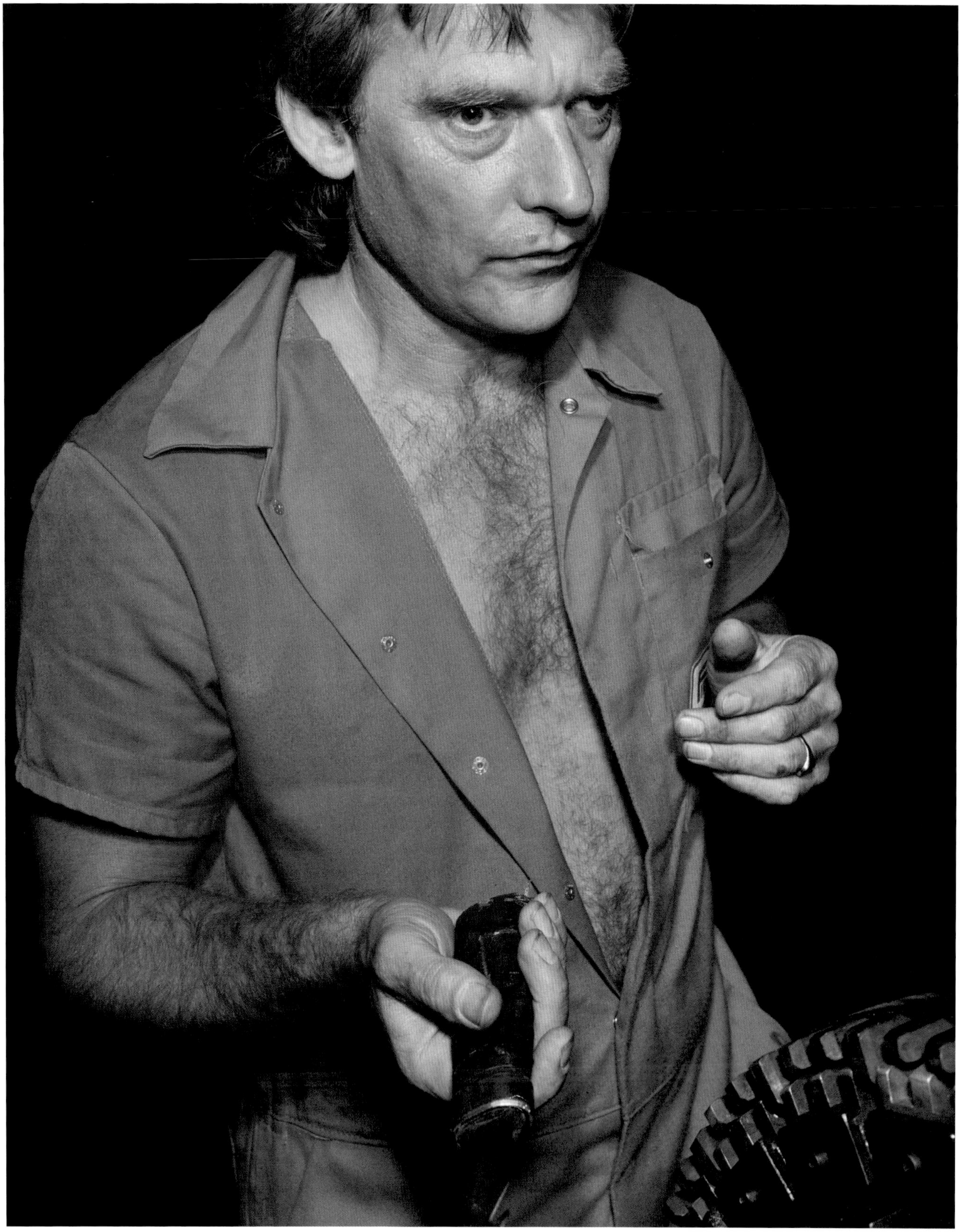

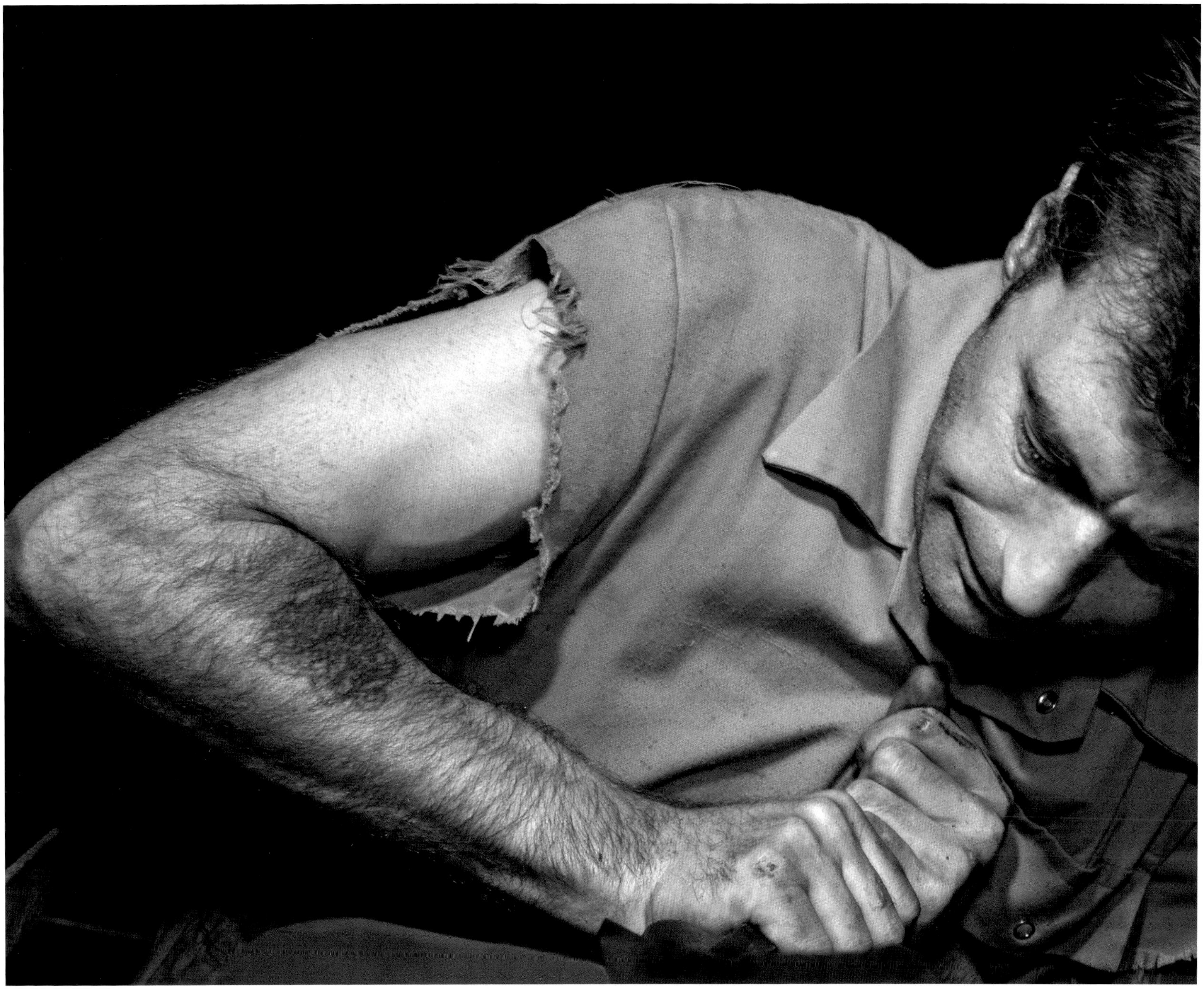

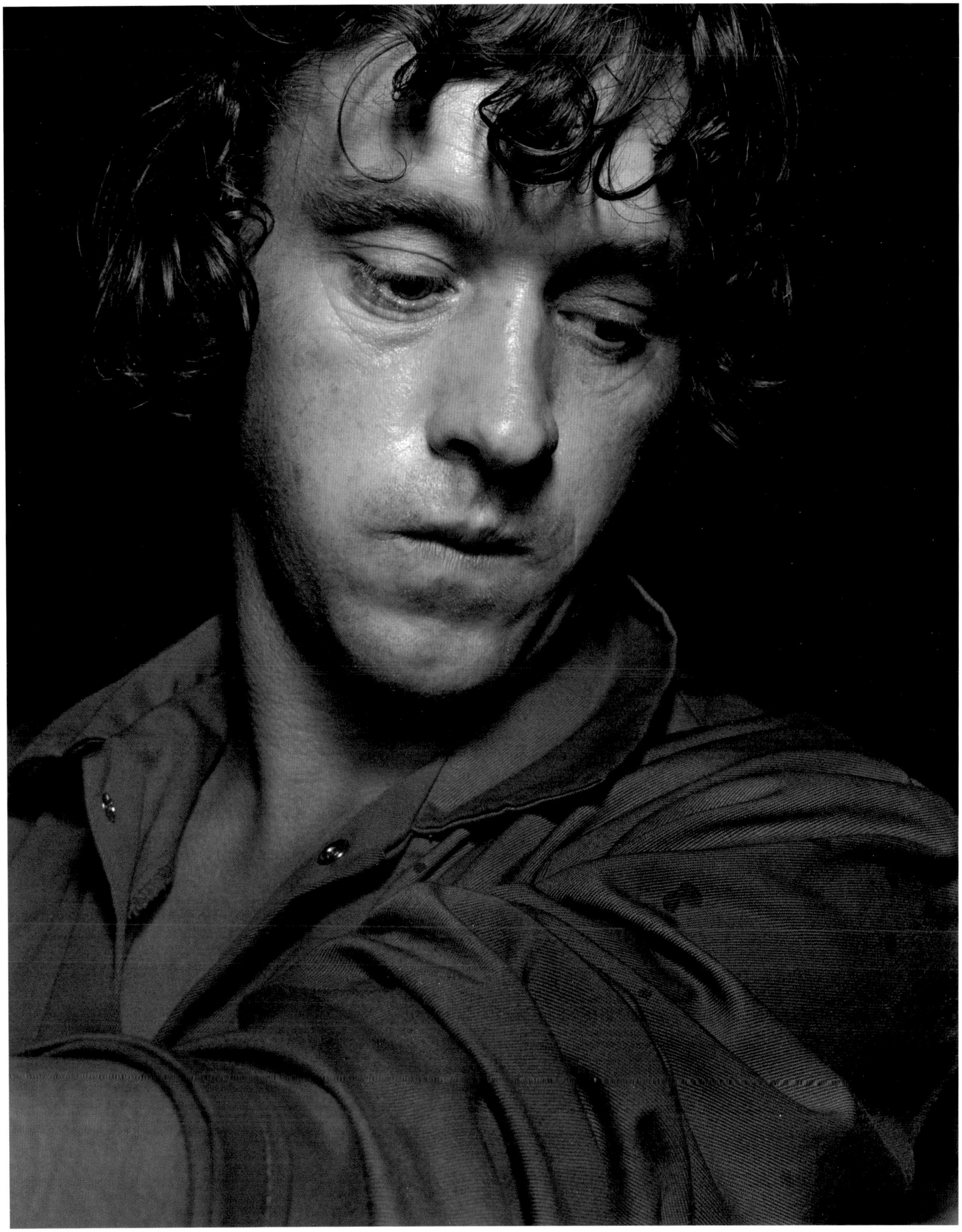

What happened
Chris Killip

On June 16 1988 I was contacted by Tom Northey, the press officer of Pirelli U.K., to see if I would have any interest in photographing in the Pirelli tire factory. I said yes, I was interested. I was also excited by this prospect as normally I had great difficulty in getting permission to photograph inside any factory. Northey explained that he had seen and liked my *In Flagrante* exhibition at the Victoria and Albert Museum in March and that he had then taken the managing director of Pirelli, U.K., Sandro Veronesi, to the exhibition. They both thought that it would be an interesting proposition to have me make a set of photographs of the Pirelli factory workforce.

It took quite some time for this commission to be formally agreed. There were no conditions imposed on me. I would have unrestricted twenty-four hour access to the factory. I could take as long as I liked to produce the work, be it three days or three months. They wanted a set of pictures, no predetermined number, which showed the workforce and was of a standard that I would be prepared to exhibit or publish. How I achieved this would be my problem. The board would only give its final approval for financing this venture if it was for a finite sum, there were to be no extras, no additional costs. We agreed upon the amount and I signed the contract on April 10 1989.

The next day, when driving back to my home in Newcastle, I decided to do a detour to Burton-on-Trent to visit the factory. I was welcomed and during my guided tour the manufacturing process was thoroughly explained to me. The factory was a vast, dark cacophony of machines, fumes and conveyor belts. The men wore dark gray or black overalls and in my eagerness to get inside a factory, I had somehow managed to forget that tires and rubber are black. I was now being told that rubber, in the processing stage, was sensitive to light, which is why all the windows were blocked out. I started to feel faint and went strangely deaf; in fact I was mortified. The problems in trying to describe this place were absolutely overwhelming. It was a photographer's worst nightmare. Why hadn't I just come to look at this factory before agreeing to do anything? If only I had, then I wouldn't have been standing there, swaying, feeling ill and foolish.

April 13 was my first day of photographing in the factory. I was using a four by five plate camera on a tripod and no lights. I had convinced myself that the only accurate way to photograph was by using the very limited light that was available. I found somewhere to live in Burton-on-Trent and I worked in the factory every day, only breaking to return to Newcastle to process the film and make contact prints. It went badly. The problems with the low light meant using a slow shutter speed. There was too much movement, not enough sharpness or depth of focus. The resulting pictures just weren't good enough.

The factory was a twenty-four hour, continuous production conveyer belt factory and it had become my nemesis. Whenever I was away from the factory I felt guilty because it was working and I was not. It was always there endlessly producing, yet it never looked any different at any time of the day or night. The only change in this cycle was the men working there on the differing shifts. I started to work such long hours that I became something akin to a permanent fixture and I knew everyone that worked in that factory. The men began to think that I had no film in the camera, that I was a fugitive in hiding, that I had flipped. At the end of six weeks I had lost eighteen pounds in weight. Tom Northey called me in late May to tell me he had just had lunch with Mark Haworth-Booth, the curator of photography from the Victoria and Albert Museum, who would now like to show the work and would December be good for me? I went back home to Newcastle at the beginning of June in order to rethink.

During the 1960's I was an assistant to different photographers in London. I was familiar with the techniques of fashion and advertising photography. It was a world that I had left behind in order to pursue my own photography. This distancing had certainly influenced my decision to use only the available daylight but it was no longer making sense. I had this repertoire of technical skills, which included lighting and surely this was the time to use them. I threw away all the work that I had done to date, except for just one image, and returned to the factory on June 14. Now without my tripod but with lights, a hand held four by five camera and the help of a friend.

I wanted to show the men in this manufacturing process as clearly as I could and to do this in this factory meant they would have to be lit. Ironically, my stubbornness in trying to avoid lighting would now have its own unexpected rewards. Because of the desperate amount of time that I had spent there, I knew in a visual way the processes of the factory intimately. The rhythms and cycles of the machines. The movements and steps that the men operating the machines had to take, the movements that the processes predetermined for them, the look of the work. I began again, rephotographing the factory with an urgent intensity using lights, sometimes three or four lights triggered by various remote control devices. The main light, the one balanced to light the subject, was often held on a pole by my friend, away from the camera, mimicking the fashion techniques that I knew from my past. I now understood and knew what I wanted to do. I embraced the artificiality of this look and its relationship to fashion, film noir, and even to Soviet realism. The workplace had become in a real sense for me, a theatre and I was now recording the perpetual reenactment of an enforced ritual with its own demanding tune.

The men knew me, I was very familiar to them; we talked about soccer, then cricket, relationships, children, politics, money and sex. Well paid, highly skilled and confident, they also had seen me so clearly circumscribed by my own machines. Tools for doing a job, all part and parcel of this thing called work. We were well rehearsed.

I finished photographing in the factory on September 13. The exhibition opened at the V&A on December 20 1989. On the afternoon of the day of the opening Sandro Veronesi came to see the pictures, on his way to the airport to attend a board meeting in Milan. It was the first time that anyone from Pirelli had seen the photographs. He slowly went around the exhibition without speaking, then after looking at the last photograph he turned, shook my hand and congratulated me on the work. I told him that I had been worried that after putting so much trust in me, he wouldn't like the photographs. He said that he liked them because they accurately showed the workforce who makes the tires and the images did not shock him. He knew what a complicated and sometimes dirty process this was. He then added, if after all this time I now knew a better way to make tires, he would be interested to learn about it.

Both Tom Northey and Sandro Veronese had wanted to make a publication of the photographs for distribution to the workforce of Pirelli U.K. although this did not prove possible at that time.

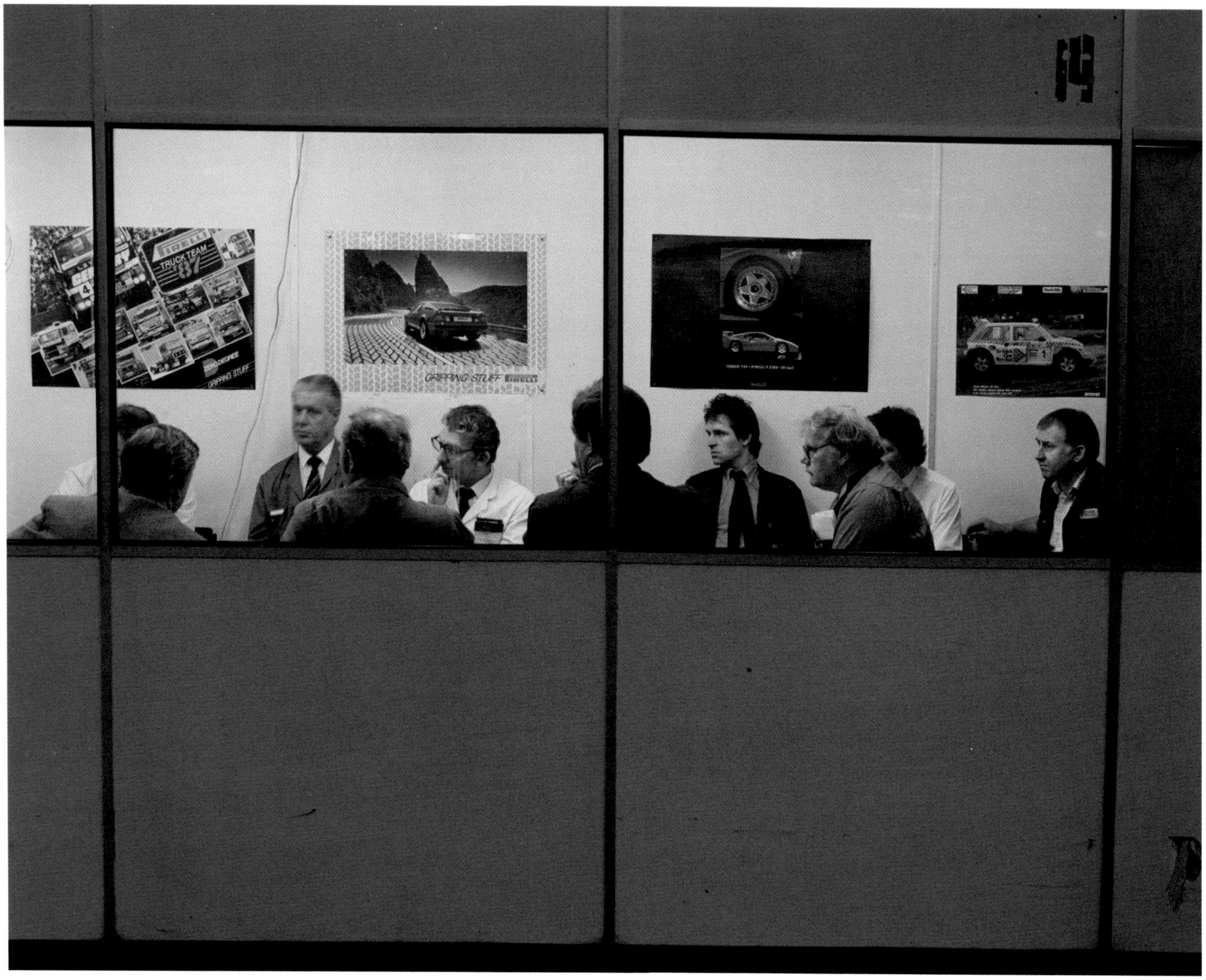
PIRELLI
TRUCK TEAM
87
ZERO DEGREE
GRIPPING STUFF
GRIPPING STUFF PIRELLI

Chris Killip's portraits of the Pirelli Workforce

Clive Dilnot

I — In a little book that has lost nothing of its critical force in the decades since its publication, the Czech novelist Milan Kundera insists that the work of the novelist—to whom the photographer is linked, and in interesting ways—is to act as a counter-weight to the forgetting of being. The novel finds its purpose, Kundera argues, in reclaiming for consciousness, 'one-by-one,' the various dimensions and possibilities of lived existence—'everything man can become, everything he's capable of'.[1] Its ethic, by contrast, lies in discovery. As Kundera puts it, in a formula that actually addresses the essence of every art, photography included, 'the role of the novel is to discover only what the novel can discover. A novel that does not discover a hitherto unknown segment of existence is immoral. Knowledge is the novel's only morality.'[2]

This reference to Kundera seems appropriate to begin these reflections on Chris Killip's *Pirelli* essay. The connection is partly to a European ethos—Killip has long been regarded as one the most "European" of British photographers (using this term in its older, more literary, sense of a density of existential probing),[3] but it is equally a matter of the issues with which Killip's essay deals. The 'unveiling, displaying, illumination' of *politically* forgotten being is as fair a definition of the essence of the *Pirelli* portraits as one could find. There is something too in Kundera's notion of work as the articulated *discovery* of hitherto unknown or un-regarded segments of existence that addresses a substantial component of what these pictures achieve.

Yet there is another peculiarity to this essay, and one that marks its character. While the *Pirelli* pictures comes into view as essentially a new project for photography, in fact the project was completed seventeen years ago.[4] There are reasons, of course, for the delay in its publication. Some will be discussed below. One irony of this situation is that it is likely that this work will have far more resonance today than when it was first produced. Nonetheless, the interval between the project's creation and its publication might make us pause. Burdened by the conceit that nothing can be written or made that can lie outside our abilities to comprehend it, we tend to forgot that our understanding of what works offer—particularly those that swim against the tide—is always more limited than we wish to admit. In fact historically we know that the evasion of the content that a work

1 Milan Kundera, *The Art of the Novel* (New York, Harper and Row, 1986) p. 42.
2 Ibid., p. 5–6.
3 I refers also to Killip's acknowledged admiration for, and incorporation into his photography of lessons learned from, figures like Ed Van der Elksen, Christer Stronholm and, especially, Josef Koudelka.
4 This essay was first published in 2006.

offers has often been the essence of criticism. In eliding this fact for our own times we reduce our thinking to what our contemporaneous position in history permits us to perceive (an understanding that is always *less* than we imagine).

Stendhal, in 1835, knew better. Exiled in Rome, determined to commence his autobiography before the complications of syphilis took his life (as they were to do six years later) but knowing also that to write for the mores and sensibilities of the contemporary French bourgeois would limit, instantly and irrevocably, the language he could deploy—*and would therefore make it impossible to tell the truths which could alone justify the enterprise in his own eyes*—he opts for a radical solution: his readers, he declares in his opening pages, will not be those of his own time but those of fifty, seventy, or even a hundred years hence.[5] To us this tactic appears extreme. But faced with the values of Louis-Phillip's ascendant bourgeoisie, and thus with the reversal of everything he had stood for (the revolution, Napoleon, 'glorious impetuosity')[6] not only did it give to Stendhal the freedom to write as he wished (hence on these grounds alone it should not be dismissed) it allowed a greater virtue—that his book could thereby present truths that the historical moment could not stomach.

There is something of this Stendhalian position in the *Pirelli* essay. True its delay is the product more of accident than intent (Pirelli had originally planned a publication of the project, but this fell through as a consequence of changes in the company). Produced in 1989, but published in full here for the first time (and aside from a single showing a few months after its completion, never exhibited again in full) the awkwardness of these pictures vis-à-vis contemporary mores in photography; the un-fashionableness of the content, and above all their refusal to accede to what the market, in 1989 as much as in 2006, wants of photography, speak to a project inimical to the time of its inception—one whose truths remain to be comprehended.

II —— Certainly, the significance of these photographs was not fully grasped when they were first seen. Their only exhibition, at the Victoria & Albert Museum (V & A Museum) in London in 1990, drew few reviews. Although a couple of those that appeared were not unsympathetic,[7] over-all both the art press and the photographic journals of the moment ignored the exhibit.[8] This is less surprising than, in retrospect, it may appear. Context, as ever, plays the major role. And the context for looking at these pictures in London in 1990 was far less sympathetic than might be imagined.

Thirteen years earlier, in 1977, *Creative Camera* had given over an entire issue to Killip's early work in the North-East of England—the first time the magazine had presented the work of a single photographer in this way.[9] At that point the work was felt to epitomize the cutting-edge ambitions of ambitious, serious political photography in Britain.[10] This still seemed largely true when, with Graham Smith, he created the important 1985 exhibition 'Another Country.'[11] By 1990 however, the situation began to change, and in quite fundamental ways. Two years earlier Killip had published and exhibited (again at the V & A museum, prior to a European tour) his major photographic project, *In Flagrante*.[12] Both book and exhibit were well received, confirming Killip's position as one of the most significant photographers in Britain. Yet even as *In Flagrante* wins for Killip the second Cartier-Bresson prize it becomes clear that attitudes, both in art and in politics, were shifting sharply away from the ethos that Killip and his work were seen to represent.

Incidental evidence supports this view. The winter 1988 edition of *Aperture* magazine, entitled *British Photography: Towards a Bigger Picture* (which was derived from a two-part exhibit of the same title held in London in 1988 and 1989), appeared as *In Flagrante* was completing its run at the V & A Museum in London, and a few months before the Pirelli portraits were exhibited. Devoted to emerging British photographers, Killip was featured in the issue, but only through a passing comment in an article by Mark Haworth-Booth ('the most important realist photographer in Britain since McCullin') and by a book review of *In Flagrante*. While the latter was respectful, like the magazine as a whole it left a clear sense that the project that Killip represented was over: 'Unlike many of his contemporaries, Killip works exclusively in black-and-white … For Killip, color has too much forced immediacy. Yet it is precisely this quality that's turned to advantage in some other current photographic work that comments on the State of the Nation, albeit in very different terms. Anna Fox's *Work Stations: Office Life in London* and Paul Reas's pictures of shopping malls and supermarkets … both approach it from the

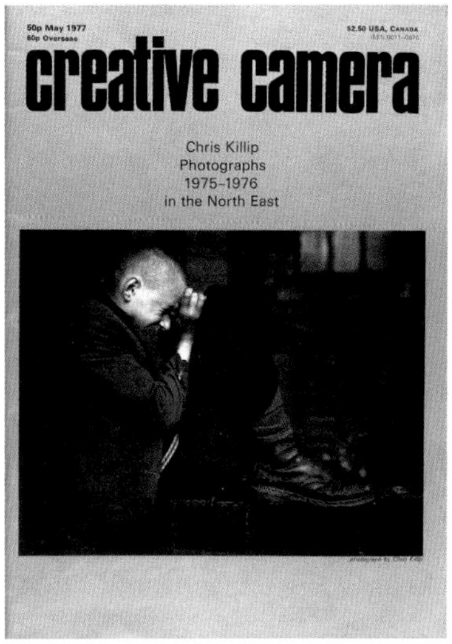

Creative Camera Magazine, May 1977

other side, the side marked off by the ascendancy of consumerism, suburbanism and a headlong aspirational pull'.[13]

This is more telling than we might assume. It chimes with the wider reaction to *In Flagrante* after 1990. In 1988 the book had been seen as the product of two decades of serious work by Killip; the culminating masterpiece of a tradition of politically-focused work. (And as the first serious "European" work in British photography). But it also seems to have been immediately sensed—and the *Aperture* review registers this—that it was, at the same time, all but a requiem; a work both magnificent (particularly given the domestic traditions of English photography) but nonetheless also a memorial, at once to a subject-matter and to way of working, and beyond that (though this is not quite yet articulated) to the wider political project it addressed.[14]

III — Given the situation of British photography of the time none of this is particularly surprising. The late 1980s and early 1990s are the moment in which, if in still relatively ill-digested ways, the implications of the digital revolution and (even more) of post-modernism become the new horizons of practice and criticism. The general embrace of photographic media by fine artists in the 1980s gave photography an opportunity to re-create itself as fine art. The last fifteen years or so have seen the working out of this development.

It was bound of course to produce critical disruption. It was after all designed to so do—to shake a field that had for far too long been too limited in its aspirations, too frightened of its own possibilities. Killip himself, at an earlier stage, was a part of this process, seeking an ambition for British photography that could take it beyond the confines of journalism and the picturesque.

But the assertion of photography's new status as fine art goes along with the creation of new critical criteria. Although unfocused and indeterminate (certainly in their implications) these conjoined the aspirations of a younger generation of photographers to be considered both fine artists and "in the critical swim" of postmodernism in general, with the needs of the museum and gallery contexts through which, increasingly, photography was seen and processed.[15] In its own way (if somewhat tritely) *Towards a Bigger Picture* lays out what this means in a series of axioms.[16] None are at all surprising. That photography

5 *The Life of Henry Brulard* was not published until the 1880s. Its wider readership has had to wait upon the current acclaim for Stendhal as one of the greatest—and perhaps most contemporary—figures of French literature.
6 The phrase is John Sturrock's. It comes from his introduction to the Penguin edition, *The Life of Henry Brulard* (London, 1995), p. viii.
7 The two significant reviews were by David Lee, in *Arts Review*, January 1990, pp 6-7 and by Janet Abrams in the *Independent* newspaper, December 29th 1989. The review by David Lee in *Arts Review* is particularly perceptive. As pre-publicity for the exhibition the Independent magazine published an article by Paul Barker ("Factory Art") on the Pirelli factory (16th December 1989). The article is useful for its references to the work processes and it reproduces six of the photographs (including images not reproduced in this publication) but it does not discuss the photographs.
8 There were no reviews in any of the photography journals. A few comments on individual pictures by Gerry Badger in *Chris Killip55* (London, Phaidon, 2001) pages 111–118 are the only other published references to the project.
9 May 1977, pages 150–171.
10 The essay is without precedent in the low-key ferociousness of its anger against what Killip could already see was the effective destruction of a culture—and this is five years before the effects of Thatcherism began to be felt.
11 At the Serpentine Gallery in London. The exhibition was largely well received, though there was criticism that the show presented an "out-of-date" 'cloth-cap" image of the North-East.
12 London, Secker and Warburg, 1988. Introduction by John Berger and Sylvia Grant.
13 Liz Heron "Real Beauty: review of In Flagrante" *Aperture* #113, Winter 1988, p. 75
14 This perception has not been helped by its relative invisibility—the book sold out its first (and only) edition rapidly and is today not easy to come by even in the second-hand market. Combined with Killip's relative reticence in publishing and exhibiting since 1990 this has tended to reinforce the idea of this as an exemplary project *of the past*.
15 A pertinent example here in regards to Killip is the "Another Country" show he made with Graham Smith in 1985. Held at the Serpentine Gallery in London, a major Arts Council venue in Hyde Park, there could hardly be greater incongruity between the content of the exhibit and its physical setting. But the incongruity is also institutional—and had the practical consequence of denying the exhibit a catalog, officials in the Arts Council feeling that the proposed, newspaper-like catalog was "inappropriate" to their (art) institutional goals.
16 The editorial view of the issue editors for *Aperture* was remarkably banal even by the standards of such surveys. The concluding lines of their opening statement read: "The picture of Britain they leave is of a country bursting with creativity, whimsy, frustration and rage, a place of familiar pleasures and implacable demands, of giddy hopes for the future combined with the nagging social and economic realities of the present.'

has become, "for better or worse," an art medium; that if one continues, despite this, to work in a "documentary" or "realist" mode it is necessary to photograph in color (photographers, *Aperture* implied, had to "get over" its immediacy, its affect); that in general suspicion must attend depiction (but latitude can be given to art) are all norms that in one way or another have become commonplace. So too the wider prevailing sense—as yet more implicit than explicit—that it is necessary in this new context to shift allegiance, if not, quite, yet, exclusively to photography-as-art, then certainly away from the photographic traditions as then understood.

But if these criteria are not surprising, it is clear, particularly as they are interpreted in the early 1990s, that Killip's pictures do not stand happily to them.[17] Looked at through these lenses, the pictures in *In Flagrante* present neither the appearance of art nor the sensibilities and conditions of the "New Britain" that was then emerging— neither its vicarious highs, nor (sufficiently) its seedy, consumerist underbelly. In terms of the new critical standards there is too much struggle in these pictures, too much on "class," too little on the ironies of aspiration, ownership and possession. It is telling in this respect that the two photographs from *In Flagrante* that are published in the final pages of *Towards a Bigger Picture* appear almost as end-note embarrassments, valedictions for a project that—at least in the mind of the editors of *Aperture*—has obviously already passed into history.

IV — If this is the critical sentiment that is deployed against *In Flagrante* it applies with re-doubled force against the *Pirelli* essay. No one had disputed, two years earlier, that the pictures shown in *In Flagrante* were distinguished by critique—if of an unfashionably empathic kind. The *Pirelli* photographs, by contrast, seem at first sight to have none of the acid bite or sharp political commentary that underlies the earlier pictures. Here are studies of work, fifty in total. All, save two, picture men. The preponderant tone is reflective; these are portraits of concentration and engagement in tasks. They are not even —this might have saved them, *pace* Friedlander—pictures of boredom (which is today the primary emotion we expect work to arouse and hence expect to see). White-collar work (office work, managerial work)—the only work a contemporary European or American audience for photography can *experientially* understand—is all but absent. A single photograph, derisively tacked onto the end "for context," suffices to deal with it (management is pushed away, seen behind glass, not a part of what is occurring on the shop floor). For the rest, the essay is unashamedly about physical work, though not necessarily as degradation, nor, pace Salgado, as romanticism.

Photographically, the references that are deployed, and they are many and various, from fashion photography to Man Ray and Germany of the 1920s to Stakhonovite Socialist-Realism, are present but muted, and therefore either scarcely registered or felt as vaguely inappropriate—as if, somehow, there is something of bad taste, something illegitimate, about inflecting these portraits in this way. On the other hand, apparent familiarity with their genre (*presupposed* as 'realist documentary') and their subject matter (understood as a variation on the portrayals of class that Killip had been developing since the early 1970s) make these photographs seem like rehearsals of the already known. Even their devotion to black-and-white and their undeniable underlying humanism—itself a sensibility under critical opprobrium—appear to speak only to issues that are in decline. To put it at its harshest, the economic irrelevance of Killip's subjects (in a generation their factory could be closed or in China, this is already sensed)[18] mirrors—or so it seems—the photographic irrelevance, vis-à-vis the future, of the black-and-white portrait.

The description is of course a caricature. But it was this caricature that was *seen* in 1990. The wider problem is that for all the regard paid to him—"a distinguished reputation and much respect through his rigorous and committed approach to a small number of extended projects"[19]—the understanding of Killip's project as a whole was beginning to slip away. But there is a reason for this, one less acknowledged in the photographic literature, but which needs bringing out. It is that the uncertainty over Killip's photography (and over the *Pirelli* project in particular) had at least as much to do with the shifting political climate of the late 1980s, early 1990's and with the desperate attempts of art and culture to accommodate to these changes.

Politics bears upon art not only directly but equally, and perhaps in the end more profoundly, indirectly through institutional pressures, conformity and adherence to powerful unspoken rules and the

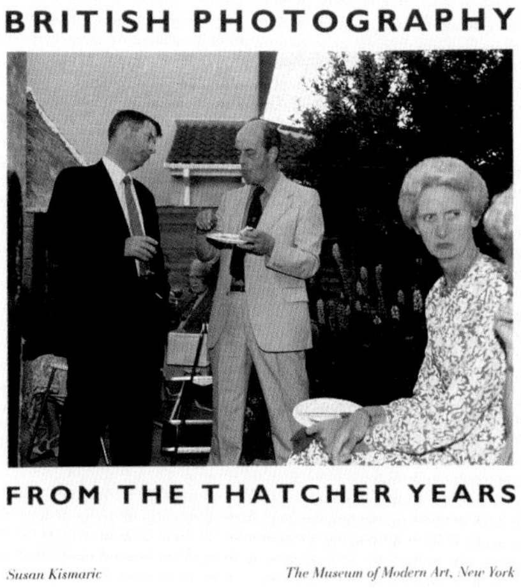

Conservative "midsummer madness" party by Martin Parr

delineation of that which it is acceptable (or not acceptable) at any moment to deal with. The late 1980s and early 1990s produced both conformism to the Thatcherite agenda and a resistance to it—but the latter took the form of cultural rather than economic resistance. In retrospect we can see that political impotence led to an over-valuation of cultural politics; an avant-gardism and "political correctness" that inevitably took the past as a whole as its enemy. In this context Killip's work *embarrasses.* No other word captures the equivocation that is obviously felt about his project, and it embarrasses because in large part because the political and social—even cultural—moment for it seems to have passed.

It is no accident perhaps that one of the two illustrations from *In Flagrante* chosen to illustrate the *Aperture* review depicts a scene from the (failed) miners strike of 1984. Without it being explicitly said the implication is nonetheless clear: that the defeat of labor at the hands of Thatcher and Reagan also marks the end of that tradition of photography which (in one way or another) took the representation of that class as its central subject-matter. Certainly in Britain, with the project of the welfare state well and truly in tatters, the idea, in the 1980s, of the working-class as the agent of history—as in *any* sense possessing a future—already seemed anachronistic.[20] To photograph labor in this moment seemed then to be almost willfully regressive. What was emerging, what demanded representation—and urgently—in this other view, was not labor but the world that conservative politics and the market were together then creating.

It is scarcely surprising then, if we turn to the key emblematic political photographs of the late 1980s and early 1990s, that they do not concern labor at all. The history that supported and secured the sense of immanent dignity as the revealed nature of human striving—the under-pinning of so much documentary photography from the 1920s through the 1950s—has disappeared. It is not, in this moment, embodied dignity (or anything like it) that holds center stage but something much nearer to its opposite, a kind of flaunting of how far degradation can go—and it is those who deal most unsparingly with this new condition who now necessarily move to the fore.

V — The photographer who, in the late 1980s, seems emblematic of these developments is Martin Parr. An artist of cynicism—probably the greatest

17 There is a graphic example of the confusions and 'crisis of confidence' engendered in this period in the catalog for a British Council exhibit "Documentary Dilemmas: Aspects of British Documentary Photography 1983–1993." Featuring Killip and a number of younger photographers/artists there is a quite startling lack of coherence in the work shown. This contrasts sharply with the (politically and aesthetically) much more focused exhibit at MOMA three years earlier, "British Photography from the Thatcher Years," which drew on a smaller number of photographers (Killip, Parr, Smith, Davies, Graham) and took a more restrictive sense of the photographic.

18 In fact the factory is still open.

19 Heron, ibid. p. 74

20 Amongst other pointers of a highly significant year for contemporary history (of which the beginning of all of the Soviet Union and therefore the global victory of capital is the most important) 1989 can also stand as the year in which what used to be called the "forward march of labor" is not only halted (as Eric Hobsbawm had it in a famous essay written after Thatcher's first victory) but *reversed.* It is a year after which, in effect, labor *no longer matters,* either for politics or for society as a whole.

in Britain since Gillray—Parr early linked himself to Thatcherism, though less as a political supporter than as the chronicler of the sensibilities that Thatcherism at once produced and fed from. The politics of the 1980s effectively made Parr as an artist. It gave him the license to discover a mode of photographic expression wholly appropriate, in its surface aspects, to the dissolution of dignity and the rise of "shoddy capitalism" that appeared in the wake of the end of labor. 'The Last Resort' (exhibited in 1986) and 'The Cost of Living' (which came out in 1989) brilliantly give form to these tendencies. The first, set in New Brighton, a declining seaside town opposite Liverpool in the Mersey estuary, treats the working-class subjects he photographs with unrepentant anthropological distance, as specimens from another world. As with a couple of earlier photographs of decaying public housing in Liverpool, taken around 1982-83, there is nothing left in these photographs either of hope or of dignity.

Killip had been cynical, in the 1970s and early 1980s about the official forms of labor—there is not a single picture of paid-work in *In Flagrante* and the pictures in *Creative Camera* essay are a sustained epitaph to every illusion of the progressive realization of a just society emerging out of decaying capitalism. But, as the epigraph to *In Flagrante* testifies ("my life does not depend on their struggle only my hopes") Killip nonetheless preserved at least a degree of hope.[21]

In Parr, all the illusions, all the aspirations, all the hopes (particularly politically), have gone. In its place is only a kind of Orwellian disgust,[22] an aesthetics (and politics) of fascinated repulsion; emancipatory, if at all, only in the ways in which the cynicism ruthlessly exposes *all* pretensions and hopes, Thatcherite as well as those still accruing to labor.[23]

Against this (at best revelatory) cynicism—for Parr rubs our noses in Thatcherism just as Killip had in the failed economics and politics of the 1970s[24]—Killip's portraits of the Pirelli workforce seem almost crafted productions. They demand a degree of interpretive encounter (and of empathy for the subjects depicted) that, in 1990, is emotionally—and politically—impossible for viewers to reach. This is why Parr's photography appears so conclusively at this moment as the photography that appears to *understand,* albeit it at a visceral level, the world that Thatcherism is bringing into being. What Parr touches on—and Killip cannot—are two conditions essential to the workings of the market: disengagement, or the removal of affect,[25] and the erasure of history.

In the end it is the latter which is the most unsettling element in Parr's photography (and the factor which by itself ensures that in relation to it a photography like Killip's cannot be seen). In Parr's pictures it is as if history is no longer present, even as a possibility—as if it is no longer that which even could be photographed. There is, as a result, a kind of terrifying closure in these pictures: the resonance of time, and therefore of past and future and their alternatives, is completely absent. There is nothing but—*there can be nothing else but*—what is in front of us; there is nowhere else for culture (for history) to go.[26]

VI — Seventeen years on, everything, and nothing, has changed. What has not changed is that the market has not disappeared—far from it. We are still playing out the consequences of the changes that occurred between 1989-91—but with this difference, *that we are no longer innocently so.* This loss of innocence is both political and photographic.

Politically, the closure that Parr emblematized—and which was expressed in that moment in at least the title of Francis Fukayama's notorious 1992 essay "The End of History"[27]— can never be literally meant (history never ends in this sense). But it describes nonetheless the sense that grips our times—the sense which issues in the expectation that we live today not progressively (the very notion is tainted) but in a kind of perpetual present, out of which, or from which, there can be no development, no change, no movement (no deliverance) and there can be none of these things because any one of them might threaten the world that Parr so well and so incipiently portrayed, that is the world converted to consumption.

But we have long since gone past the small (ersatz) exhilarations of the market just as we have gone past the allure of what it seemed to offer as a subject for representation. *Then,* there could seem to be a value in finding a photographic language that was objectively capable of depicting society treated as a market. Today, this work looks less mimetic (in the strong sense of the term) than merely naturalistic; the depiction of what-is presented in the latter's own terms. The flattening of the world, which seemed of such consequence in terms of (photographically) grasping the early stages of the market's global victory—and Parr was as prescient in reading these developments as Killip was in picking up the politics of exclusion and disposal in the 1970s—now looks closer to capitulation to those same forces. In fact, since 1990 there has been a recession in critical photography. Boris Mikhailov in Russia—who has been producing an extra-ordinary body of work since the 1960s[28]—may be the last great exponent of trenchant criticism.[29]

VII — Where does this leave us? Specifically, to bring us back to the *Pirelli* essay, how do we make it possible to *see* this essay, not as anachronism but as possibility?

This is not easy. There is a way in which what these photographs picture—or at least what they *seem* to picture—has already disappeared for us (meaning disappeared for us as part of our experience). These photographs seem part of that history of the disappearance from our economies of work in the traditional sense. But that disappearance is not simply historical, not just a result of "blind" objective forces; it is political, and economic, a result of decisions-made—and less blindly than is still fondly imagined. Especially in Britain, the *deliberate* destruction of those parts of the economy in which labor had power (mining is the obvious but by no means the only example) was a clear political (and economic) project. There is a line in Marx that is relevant here. It comes towards the very end of *Capital,* where he is talking about primitive accumulation and its transformation into early capitalism. He notes 'These methods [by which these transformations occur] depend in part on brute force .... Force is the mid-wife of every old society which is pregnant with a new one, it is itself an economic power.'[30] We have tended to forget the extent to which behind 'market forces' lies force—which in our economies means the capacity to enforce or compel increased inequality—and hence insecurity. This in itself suggests a greater need to contest the implications of what is contained in the victory of market forces; to think this victory not only culturally but through talking again about inequality and its implications. In that sense economics needs to come back onto the *cultural* agenda. But this might also remind us that the disappearance of labor is not an

21 In this respect Killip is still acting roughly in accord with Richard Rorty's salient summary: that one's efforts should be devoted to the creation of a social democracy that 'is a classless, casteless, egalitarian society.' If the chances of realizing such are 'pretty dim' this is 'not a reason to change our political goal. There is no more worthy project at hand, we have nothing better to do with our lives.' See Richard Rorty, *Philosophy and Social Hope,* (London, Penguin, 1999), p. xiv.

22 The reference is to *The Road to Wigan Pier* and George Orwell's journalists trick of seeking out the worst possible places to inhabit and describe. Orwell's middle-class audiences, left and right, swallowed the bait whole. Those who knew the context despised the conceit.

23 To Parr's credit, "The Cost of Living" applies an equally distanced and disengaged eye to the middle classes. Skewering the mentalities that that brought Thatcherism into being "Conservative "Midsummer Madness" party, Bath, Avon" is by some way the single photograph of these years that most powerfully captures the cynicism, venality, complacency (and underlying aggression) of Thatcherite politics—tellingly MOMA took it as the cover image for the catalog of "British Photography From the Thatcher Years."

24 Parr's photography gains its force precisely from its wholly unsparing disengagement from *any* commitment. In so doing it offers, through the distance that irony creates, both a vicarious exhilaration (and, through its own legitimation as commentary) a degree of absolution for the perverse enjoyment derived from the degradation to which subjects in a market culture—"them" not "us"—can be reduced. Parr's genius at this point was not to compromise in any way on this vision—thus providing in his own resoluteness a parallel to the determination with which the market ideal was stamped onto Britain.

25 We tend to forget, although we should not, for it reaches into our times, that, after 1979, and more especially after the mid-1980s, political (and with it moral) disengagement became for many (especially in the middle-classes) almost the only tolerable political philosophy. Disengagement was a way of dealing with impotence, a way of forgetting ("because nothing could be done about it") the implications of policies that, however socially repugnant, allowed for many a welcome relief from political thought.

26 The photographic agent of this disappearance is color—but used in a way that pulls depth out of the picture. The result is that affect and resonance vanish, to be replaced by a kind of blank immediacy, an emblematic literalism that makes it appear as it the world is self-representing itself in its banality.

27 *The End of History and the Last Man* (New York, Avon Books, 1992). It is interesting that Fukayama, having begun as an enthusiastic neo-conservative, is now beginning to reverse himself, particularly in regard to the neo-conservative movement's espousal and drive for the Iraq war. Is there therefore once again history?

28 Summarized in *Boris Mikhailov, A Retrospective* (Zurich, Fotomuseum Winterthur, 2003).

29 It is clear too that the romance of photography and fine art has long since past its honeymoon stage. The liberating moments of this relation belong exclusively to the 1980s—and few, if any, originated out of photography (photography proving of far greater value to fine art than has fine art to photography).

30 Karl Marx, *Capital* (Harmondsworth, Penguin, 1976) p. 916

inevitable Darwinian process—it is a resolute political act. To be sure, the market has persuaded nearly everyone that it is (paradoxically) in their interests to "give up work" in the traditional sense—to attain to consumption through renouncing labor in favor of 'flexible' participation in the service economy. Today however, even orthodox economists are smelling a rat.[31]

This is not irrelevant photographically speaking. As we've seen in Parr, the erasure of history in the market is the erasure not just of past and future but also of the depth—the resonance—of the present. It is coupled—we can all feel this—with the transformation of subjects who *were historical actors* into figures, personifications, ciphers. This was ever a problem. No less for labor than for any other subject. The photograph has always been a prime agent in this process. Its capacity to reduce is terrifyingly powerful. But photography has also—and this we badly need to recover—been a source of the opposite, that is of the bringing of subjects into a view as political subjects, as subjects *for* politics.

But to do this in relation to photography today depends on prior work to recall what it means to represent or to depict with, or on behalf of, a social agenda—and that in turn seems to necessitate not just recovering the idea of the critical project of the good society (and once again believing that such might be a possibility worth struggling for) but also a notion of the democratic political subject—of subjectivity as that which *includes the political within it* as a part of the *minimum* definition of who a subject is.

VIII —— There is, in Gadamer, a very beautiful line that helps to explain what is at stake here. It comes towards the end of a section of *Truth and Method* in which he is exploring, in some depth, the work that pictures do. Having noted that, 'in representation, the presence of what is represented is *completed*' he adds, a few pages later: 'Word and picture are not mere imitative illustrations, *but allow what they represent to be for the first time what it is*'[32] (my emphases). Politically, we can interpret this statement in the following way: that picturing it is not (only) a matter of representing the subject who already is a political subject, it is rather that the *act* of representation, of picturing the subject, *itself* brings the subject to view as, potentially or in fact, a political subject. It is therefore representation that makes ('completes') the subject. Or, to put another way, in a sense that immediately resonates to the history of photography, it is picturing that allows 'for the first time' the subject represented to be what he or she is, and what he or she would wish to be, i.e., a political subject. Representation then, in this affirmative sense, is the act of *making* the subject political.[33] It is the insistence, *in the photograph,* that here is a subject *to whom must be accorded political rights.*[34] This is the un-paralleled political insight that justifies, for example, despite all else, FSA photography. In an epoch in which representation is dominant, *not* to be represented, *not* to be shown, is not to have a voice—it is to be politically excluded. Conversely, to bring subjects to visibility, such that they can have a voice in their history, such they can become subjects in and of history, is a political task. At its political best photography uncovers overlooked dignity and in so doing anticipates a democracy to come. It is the anticipatory discovery of this essential political dignity of the overlooked (today: the excluded) that is the mark of politically progressive photography—and it is the triumph of this photography, from Marion Post Wolcott to David Goldblatt, to make this dignity credible.

The *Pirelli* essay lives within this orbit, but to recover its meaning for today we have to see this project not regressively or nostalgically, not (only) as history, not of then ("then" being the moment when the traditional or historical project of labor still seemed alive) but as anticipation—*and of what is to come.* "What is to come" are the impossible consequences of globalization. In the ruins of the globalization that can never be completed (because the forces that it unleashes are ultimately unsustainable) are both the crises (of economics, of ecological depredation and consequence, of demographic disaster, of the loss of any pretence at democracy) that we will have to live through, and the roots of what will, in its myriad and different forms, have to replace incoherent and dysfunctional pure-market economies (John Gray). In these spaces both exclusion and labor take on new meanings—the one as the anticipation of that which will become increasingly ubiquitous; the second as that which 'comes back" not as labor (in the classic sense) but as the *act of producing.* Irresponsibility towards production is not the least of the sins of irresponsible markets. Conversely, *responsibility* vis-à-vis production (and hence responsibility to, for those who produce) will be a mark of what seeks to replace deficient capitalism. But what is it to produce? The historical significance of the *Pirelli* project is that it begins to explore this question.

To see this requires a re-construction, ideally first of Killip's project as a whole—for these pictures are in many ways the "completion of a cycle" of Killip's photography ("the last and very necessary piece of in the jigsaw puzzle of my work in Britain"),[35] and then of the Pirelli project itself. The former cannot (for reasons of space) be attempted here. But the second can. In that sense one can see these pictures through two lenses—*retrospectively,* they 'complete" Killip's work in England; *prospectively,* they open another project, one that is larger than has been previously ascribed, and which breaks with the geographic or cultural limitations of Britain.

Part Two

IX — Given that *In Flagrante* was essentially a collection of photographs of lives *outside of* or at the edge of work, nothing might be thought more surprising than that Killip should be invited by the Pirelli Company to photograph in their factories in England and to produce a body of work "documenting" (their word, not Killip's) the work force. How developed Pirelli's interest was in such project before they encountered Killip's work—or whether, as it seems, the project arose as a response to a possibility they saw in the *In Flagrante* exhibition, is not fully known. Given that the major actors are either dead or have left the company, it is difficult (in the end perhaps of little account) to reconstruct exactly the impulse that led Pirelli to decide to commission Killip.

As in all such commissions an element of vanity was surely present. At minimum, the project reflects a robust confidence on the part of Pirelli's management at that time, a sense that what they were doing—in their view, *making* tires *well* (both emphases are significant)—*deserved* recording[36]; and Killip, because, rather than despite, *In Flagrante,* is the photographer who could "do them justice." The acclaim that year for *In Flagrante;* his inclusion—with laudatory comment ("....a view of England which is the most eloquent seen in photography in Britain since Bill Brandt")—in the Royal Academy's anniversary survey of 150

31 See, as an indicative straw in the wind, John Gray's discussion of a number of recent books on the global economy, "The Global Delusion," *New York Review of Books,* April 27th 2006 pp 20–23, especially page 22.

32 Hans-Georg Gadamer, *Truth and Method* (London, Sheen and Ward, 1980), p. 122, 126.

33 This is the opposite of representation as we have thought it in the past decades. Work's like Edward Said's *Orientalism* (1978) brilliantly exposed the structures of political representation employed to *deny* subjects political identity. Today, having grasped these arguments, we need the reverse—the articulation of how, in representation, subjects can be saved from what capital would increasingly like to do to them, i.e., to exclude them as political subjects.

34 This also means: bringing subjects into the same plane as us, for politics (dialogue) does not happen with the other (the other is always merely used, or disposed of—is *not* a political subject) but only with subjects who are formally symmetrical to us. It is only through establishing relations of symmetry vis-à-vis the other that an ethics and a politics that are genuinely democratic can be established. The political issue is then that of the *same,* not that of the other.

35 Killip, note to the author.

36 There seems at that time to have been something of an almost nineteenth-century pride in their factory in Pirelli—in themselves as a company (a high-minded ethos still pervades their corporate mission); in their product (ubiquitous, mundane, but nonetheless essential); and above all in their management of the workforce. The latter, under difficult, and in a real sense more or less invisible conditions, labored with high levels of skill and productivity to make Pirelli UK an extremely profitable part of Pirelli's worldwide operations: 'We are not a poor company' as was said to Killip in the negotiations for the commission.

years of photography, and, for good measure, the announcement that he is the winner of the second Cartier-Bresson award for photography, all make Killip, from Pirelli's point of view, a bankable asset. They can rely on his expertise.

But *In Flagrante* is also an essay photographed from *within* the working-class. If Pirelli were seriously looking for a documentation of the workforce that would not be seen as merely a public-relations exercise, then Killip was their natural choice.[37] Moreover, there are technical reasons to go with Killip. Effectively closed-off from the world in many of its aspects, the plant is an extremely limited locale presenting a heavily restricted range of pictorial opportunities (all the more so given the very limited levels of natural light within the factory). Killip had particular expertise in projects developed in limited locales—*In Flagrante* was a demonstration of it, as was his earlier book on the Manx.[38] Pirelli seems to have anticipated from the beginning that this would be an extended project. Killip was available for this and supportive of it being so—indeed, given his aversion to photojournalism he would not have taken the job on any other basis. It was integral to his approach that what would emerge from the commission would do so as a result of a sustained involvement: Pirelli's interests in a substantial project, offering unlimited access for as long as was needed for the commission to be completed, would allow this to occur.

The attractions of the project from Killip's side were equally clear. In the Northeast in the late70s and early 80s, he had made various attempts to photograph inside industrial plants but had always been rebuffed. The Pirelli commission was therefore a significant breakthrough and Killip is clear about this: "The prize that Pirelli offered was the thing itself. This thing called work. I wanted these photographs to be a dissection, an examination, an uncovering and ultimately a confrontation: a showdown. To show work, revealed through those who do it, who are affected by it; made and unmade by it."[39] Work was the attraction then but so were the circumstances that Pirelli offered. Although they talked about "documenting" the factory it was clearly not a disinterested documentation of technical processes that they had in mind. The stress was firmly on the workforce. This was consonant with Killip's interests and ambitions. If the Pirelli commission would be a project with its own agenda and possibilities, it would nonetheless be close enough in spirit to extend and echo, as well to complement, *In Flagrante*. The interests of Killip and Pirelli then, at least at this stage, converged—as they did also over the issue of work.

X — For both parties this question was central. To Pirelli it was the source of their pride in their factory and their workforce (meaning: their management of it). For Killip, it was that which he most wanted to explore—that which would enable him to "fill in the jigsaw" of his photography in Britain since 1973-1974.[40]

As becomes clear from even a cursory study of the photographs, the processes involved in making tires are more technically complex than might be imagined.[41] Rubber is neither an easy, nor clean, material to handle. In 1989, the techniques and processes in Pirelli's Burton factory combined sophisticated automated and numerically controlled machinery with almost nineteenth century ways of handling and working the base-material. One result of this is that there is a much greater preponderance of complex processes demanding considerable levels of skill and concentration than might be casually supposed.[42] As significant for what emerges in these photographs is the organization of production within the plant. In 1989 this is not as what we today expect, i.e., the assembly line, but a more complex flow of material and partly-formed product that is moved from process-to-process, the stages controlled largely on a one-man-to-a-process basis, with each man working in relative autonomy and often in some isolation from others in the factory. The result, from a photographic perspective, is that the moments of work in the Burton factory can be grasped as a series of individual tableau. It is these solitary theaters of work, Killip quickly realizes, that are the potential occasions for his photography.

It is the pictures of these moments that constitute the core of the essay. It is striking that in an essay on work, with the exception of a group picture of management (seen, tellingly, a distance, behind glass, from across the shop floor) only one picture in the series shows more than a single figure (and that one is essentially a companion piece). Indeed, aside from a few photographs that show something of the machinery, and give a limited sense of context, *all* of the rest, well more than two-thirds of the total, focus on single figures. In all but two or three of these photographs the figure is at work, not rest. These are pictures of concentration, involvement and *effort;* pictures about dealing with the demands and rhythms of work, materials and processes. They are in short pictures of the actuality of shop-floor work.

XI — But the moment this is said, a qualifier must be inserted. If it is work that is in front of the camera; if it is work that is photographed; if it is work that at first we see (in 1990 it was *all* that could be seen, which was one reason for the difficulty of seeing these pictures), nonetheless, what results are not simply, or even perhaps at all, photographs of work. It takes a little time with the photographs before this becomes apparent.

The first clue that work is not the real subject of these pictures is that in very few of these pictures (in less than half a dozen) is the product recognizable as such. In a couple of the images tires are being handled, in one a hand appears to be lifting a tire-rim, that is about the extent of "product." This is significant. It does not only mean that photographically that there is little or nothing of importance that the product reveals (ubiquitous, necessary, but of little significance in themselves). Much more important is the way that the refusal to deal with product is a way of distancing the *act* from the specificity of its occasion—which *happens,* here, to be Pirelli and tires. It is a way of saying that the product does not give meaning to the act; that it has its own meanings, that it carries its own weight, above and beyond "product."

The second clue that work per se is not really the subject of these photographs is given by the simple fact that in nearly all the pictures that show work-in-action—and this is the very large majority—the exact processes underway remain largely opaque. Close study will let the viewer hazard a guess at one or two. But lacking direct knowledge the photographs do not—that is they *choose* not—to inform us further.[43] So these are not, in any sense, *industrial* photographs. (There is, in fact, across the project as a whole, an almost complete absence—in terms of approach, method and result—of any documentary intent).[44] The review of the *Pirelli* exhibit that noted that from it one could learn nothing about making tires was wholly correct in its complaint.[45]

But these are also not photographs of work in

37 It may be of relevance here that Pirelli had acquired in 1980s both welcome publicity and unwelcome notoriety for their series of explicit soft-core annual calendars. A good deal of cynicism attended their name around issues of the image.

38 *Isle of Man: A Book about the Manx* (London, Arts Council of Great Britain and Zwemmer, 1980).

39 Note to the author.

40 Killip's *Pirelli* essay cannot be understood without looking at the trajectory of his work as a whole, from the Manx pictures of the early 1970s, through to his photography in the north-east in the 1970s (represented in the essay for *Creative Camera* of 1975) and the particular foci on the "Seacoal" (Lynemouth) and Skinningrove pictures in the 1980s—the pictures that make up the core of *In Flagrante.* Space restrictions have made it impossible to deal with this is this essay. Some sense of his trajectory can be given in *Chris Killip 55* op. cit.

41 See the article by Paul Barker referred to at note 6 above.

42 There were two Pirelli factories in the UK at the time. The second, and smaller, in Carlisle, was much more automated than the Burton factory. Killip chose the Burton factory to photograph precisely for the fact that the processes employed in it still demanded high levels of skill and control from the workforce.

43 This is particularly true of the photographs that make up the present essay. In Barker's article in *The Independent* referred to above (note 6), the descriptions of the processes are accompanied by photographs that are more illustrative of the work process as a whole. In the 1990 V&A exhibition, the photographs were laid out in terms of the production sequence. In the essay presented here all attempts at portraying the work process have been abandoned.

44 These pictures do not document because to document means that there is a defined x or y which could be recorded, a body of evidence that could be collected and shown. But these photographs make no forensic claims of that order. Considered outside of the subject shown they have nil evidential value. Neither product nor process is therefore "on view."

45 This has direct implication for what the project *is.* It suggests most obviously—but this is anyway in line with we would expect from Killip—that the project is an essay not a treatise. It tells us too that if the project consciously refuses the mantle of 'documenting' then this is in order to allow the photography to explore (to discover) what else pictures of work and pre-eminently pictures of subjects at work can reveal. In other words this tells us that for these pictures work is an occasion: their goal is the exploration of what picturing work can lead to us understand, what can be revealed *through* it, even, perhaps, in spite of it.

a third sense, and that is because it is in any case all but impossible to photograph work in ways that is adequate to its *lived* experience. Work unfolds over time. No matter how repetitive the actions, *experientially* work cannot be broken down to its moments. Duration (continuity, repetition, variation, adaptation) is the essence of work, and hence *enduring* (some times with pleasure and interest, more often with boredom, not infrequently with pain and exhaustion) is its subjective condition. Photography, whose field is the moment, is by definition poor at dealing with duration. If, through the moment seized, the photograph attempts to encompass duration—as, in a sense, every photograph that tries to describe experience must—the limits of the inferences we can make from it are immediately tested the moment we begin to interrogate any photograph as to the nature of the experience (s) attached to it.[46]

It is this limit—it is all but absolute—that explains why photographs of work fail as such, why they are never more than pictures *of scenes of work*. An early photograph by Killip makes the point. *Threshing, Grenaby, Isle of Man* (1972)[47] is a very beautiful image of a now vanished agricultural activity. Yet as a record it tells us less about the nature of agricultural work (and particularly about labor) than might be supposed. The inference needed to move from the photographic scene to the durational reality of work is beyond that which the photograph can contain.[48] Photographers have long known this. The most effective photographs of work impute the conditions of work—and especially its consequences and implications—through the body. Both Evans and Strand show the body in this way, and so, if differently, does Bill Brandt. In his famous photograph of two maids waiting to serve dinner,[49] their pose—hands held dutifully behind starched apron, heads bowed— embodies the discipline that (just) contained class-relations in pre-WW II Britain.[50]

XII — The last example indicates why there is yet a fourth sense in which the *Pirelli* pictures are not simply pictures of work. The photography of work, even in photography with an avowed social agenda, has a tendency to erode the subjectivity of the subject who is pictured at work. But the *Pirelli* pictures are not reductive in this way. This alone indicates their interest. On the contrary, in them, as we will see, the subject *of* the picture is the subject *for* the picture. One reason that Brandt's picture of the two maids is so successful is that in this picture Brandt manages to suggest—however minimally, however fugitively—a concern for the lives of his subjects (a register of their existence) that goes beyond simply their use to him as the occasion for his picture.

In pictures of work this is surprisingly rare. In Friedlander's *At Work*[51] for example—a book which makes use of work as a casual subject matter without serious social intent—his subjects "are" only in so far as they are at work: they do not figure as subjects for-themselves. In Brandt, by contrast, the minimal recognition that he gives the two maids of their existence as subjects *beyond and outside the roles they are playing in work,* is sufficient to change the disposition of the picture (and to charge it with quite different meaning). The picture is now doubly motivated; it is at once a commentary (on a way of life) but it is also, even if minimally, a recovery of a subjectivity, of an existence that lives beyond the types into which the subject is cast, in life as in the photograph.

This relates directly to the *Pirelli* pictures. One of the problems that all photography encounters in general is that Barthes' seemingly common-sense dictum that 'in the Photograph the event is never transcended for the sake of something else'[52] is true only as an idea. *In practice, the reverse applies.* In the vast majority of photographs taken in the world, commercial and private, the situation, the subject, is *always* transcended, at least to a degree, for the sake of some other purpose.[53] Pressed into another's service, the subject is always thereby reduced (even beyond the reduction that the photograph in any case induces). To say the subject is made to conform (is domesticated) is to say that he or she is personified, i.e., that he or she is transformed from a concrete subject with particular identity into an abstracted figure, an object whose role is persuasive or identificatory—*typological* in the sense of providing an example. This is particularly an ever-present danger in photojournalism. The felt need to communicate to, and above all *with,* an audience's—which means with an audiences *existing* levels of understanding—tends to occasion, even unconsciously, a reversion to stereotype—all the more insidious because it appears in photographic garb and therefore takes on the mantle of the real.

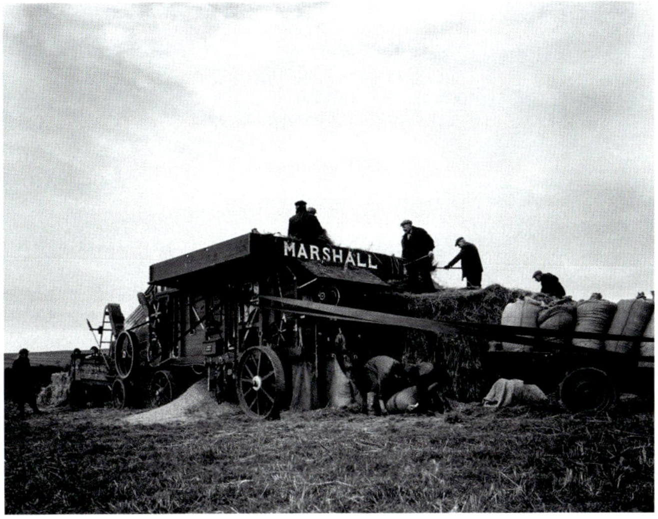

Thrashing Grenaby *Isle of Man,* Chris Killip, 1980.

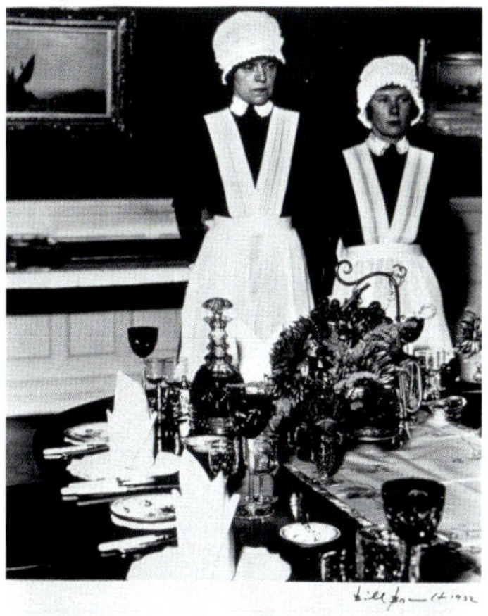

Parlourmaid and under-palourmaid,
by Bill Brandt.

46 David Goldblatt, with his sometime insistence on the extensive caption to his photographs, is one of the few photographers willing to openly acknowledge the limits to information and experience that can be coded into the photograph.
47 Reproduced on page 26 of *Chris Killip55.*
48 Another example makes the point. The least informative of David Goldblatt's South African photographs are those attempting to show the conditions of labor in the gold mines. Salgado's over-dramatized and rhetorical exercises are another case in point.
49 'The Perfect Parlourmaid', *Picture Post,* 29 July 1939. (Reproduced in Ian Jeffrey, ed., *Bill Brandt, Photographs 1928–1983* (London, Thames and Hudson, 1983).
50 It is interesting in this respect that it takes film—for example Robert Altman's *Gosford Park*—to give anything like an adequate sense of what exactly was entailed in work as a footman or parlour-maid in the world that Brandt shows. Nonetheless, Brandt's picture captures for perception an essential reality: the tension between ascribed role ("parlour maid") and the subject "within" and beyond that role.
51 New York, DAP/Distributed Art Publishers, 2002.
52 Roland Barthes, *Camera Lucida* (new York, Hill and Wang, 1981) p. 4
53 Is this point of transcendence the problem of ethics? Alain Badiou, in a provocative essay has characterized the situation and its understanding as the base of all particular ethical actions: "Thesis 3: All humanity has its root in identification in thought of singular situations. There is no ethics in general. There are only—eventually—ethics of processes by which we treat the possibilities of a situation." Ethics: *An Essay on the Understanding of Evil* (London, Verso, 2001) p. 16.

In W Eugene Smith's famous photograph of *Three Generations of Welsh Miners,* 1950, three miners, apparently returning from work (blackened faces) turn to look quizzically at the photographer, (blackened) row-houses and landscape behind. There is nothing, of course, inaccurate about the photograph. Both miners and landscape existed in just this form. And their bodies (and particularly their faces) embody precisely what we imagine their experience to have been. But that is precisely the point. The photograph is less a picture of "three Welsh miners" than a picture of what (in 1950) it was already imagined (already presupposed) three Welsh miners returning from work *ought* to look like. Smith's work beautifully cements expectation.

The political problem of such photographs is that they lead nowhere.[54] They permit no 'development' as Barthes would say (and would, in this instance, be right).[55] Instead of questioning a given identity and getting beyond its limits, the picture (even if unconsciously) repeats the typology. In so doing it condemns those who are so depicted to the limits of that assigned role: it confines them, not to say disciplines them; it gives them, at extreme, the limits of their lives (something that conservative interests are always interested in achieving). The *Pirelli* pictures by contrast stand against this sort of identification. Their focus is the subject. Like Brandt's photograph of the maids, they are an attempt at a recovery of the subject *from* simple identification of a subject with a type or a role—even as they deal, as directly as possible with that role.

XIII — What Killip understands is that to photograph work *well* means to rescue both work and the subject-who-works from the limits of existent categories. From the side of the subject this means to contest what happens through work—to contest that identity is *simply* determined through the nature of the work one does, and to contest the idea that work and its typologies can subsume a human identity in this way. From the side of work it means to refuse its false dominance. But in addition it is to seek to give weight to what else (besides boredom, endurance, irritation, frustration, exhaustion and all that demeans and denies) also happens through work. To *think* work photographically is to think again about the transitive dimensions of the experience—that which Barthes famously claimed *resisted* myth[56]—while attempting refusing reductive identifications and typologies, all that surrounds the myth of work, whether on the right or the left.

These tasks can be given specific content. One problem with the vast majority of the pictures of work is that through them nothing is *discovered* concerning it. Work is banal, frequently boring, sometimes complex; subjects at work are sometimes attentive, sometimes distracted, frequently bored, often exhausted, occasionally in danger. All this is known. But the act of photographing *the known* adds nothing to understanding. It flatters (given) understanding: 'commonsense' is (once again) vindicated. But work itself stays at a remove, *illustration in no way becomes revelation.* The first job of a (critical) photography of work therefore is to create the conditions for discovering "what else" lies in work. This is a matter of vision, of seeing, but since seeing is a product of conception (of the categorical lens through which we see the world) it is equally a matter of concepts and categories. This is difficult. It requires, as a first step, refusing the *meaning* of work in where we assume it lies—in what is produced, or in the act of labor as such.

We have already seen that Killip eschews product—for to reduce labor to product is to reduce subjects to agents who facilitate output. It is interesting in this respect that the two or three pictures in which the product is made visible are also the most obvious pictures of pure labor in the series. By contrast, and in a reversal of Hannah Arendt's axiom, in the pictures that show *work* the product disappears.[57]

But labor too allows no valorization: it has, for Killip, no intrinsic value. Labor is labor. It can be registered, and on occasion it is important to do so. To remember the physical (or mental) effort involved in making is not irrelevant—for work etches itself on the body (and in the mind). It takes a toll, and that is worth remembering—particularly for a population that have almost lost any notion of what it means to do physical labor. There are a small number of photographs in the Pirelli essay that perform this role—the essay closes on one such sequence—and they are valuable because of it. But labor is not, of itself, a source of meaning, and to acquiesce to the ideology that it is, is again to court reduction (if, this time from the "left" rather than the "right").

But if it is neither labor nor product that can

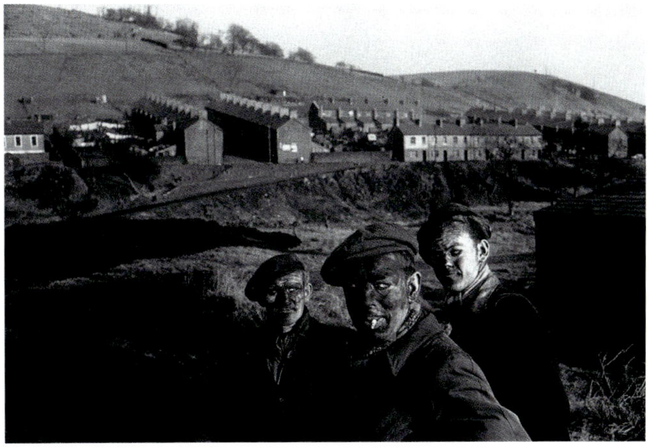

Three generations of Welsh miners, Eugene Smith, 1950
from: Gilles Mora / John T. Hill, W. Eugene Smith. Photograph 1934–1975., Paris / New York. 1998, p. 95.

be pictured, what else could be brought to visibility through work? One possible answer is that the actions involved *in* work are sensed as revelatory of aspects of existence that go beyond output and labor. The gambit has to be—here we return perhaps to the early Marx—that work provides the occasion for opening us up to some hitherto relatively unsuspected or overlooked aspects of being; further, that the subject in work can be revelatory of aspects of subjectivity in general and of political subjectivity in particular.

It is perhaps worth recalling here Elaine Scarry's sentence in *The Body in Pain* that 'achieving an understanding of political justice may require that we first arrive at an understanding of making and unmaking.'[58] Scarry's line, disgracefully un-thought of in all contemporary theories of political justice, reminds of an insistent connection, one of whose dimensions, and not the least, is the question of capabilities for capabilities (for which also read potentiality) is the heart of both social justice and making.[59]

XIV — But that work could be a source of subjective revelation runs as counter to the general run of the contemporary photography of work as it does to philosophy. The fashion—and *At Work* is again the epitome of this—is to reduce work, to abstract it. Superficially, the blankness of Freidlander's pictures appears to mimic the blankness of work. But work is not only blank in this way. And to picture it as such is in reality less to present a critique of work (nothing easier to portray than the dumb stupidity of work) than it is to *run the risk of reducing those who work.* "Where the world of the object is not taken seriously neither can that of the subject": Lukács' warning remains. Remove the density of work; remove exploration of the subject's necessarily complex engagement *with* work and the subject vanishes with it.

Today, even those in white-collar occupations are discovering this truth. Work reduced to the blankness of output recoils not on these who dictate such work but on those who undertake it. Thus not to reduce those who work—which means, in the end, all of us save those today denied access to work—means not to reduce work itself. The alternative is to acquiesce in what language already tells us happens to those who labor, that is that they are demeaned ("hands"

54 The picture could be said to come surprisingly close to the Kundera's definition of kitsch as the reverse of discovery—the desire "to gaze into the mirror of the beautifying lie and to be moved to tears of gratification at one's reflection." See Kundera, *ibid.* p. 40–41
55 The reference is to Barthes' argument (in *Camera Lucida*) that the photograph can *never* be the motor of thought, i.e. development. This argument is, in principle, wrong. It is in this instance however correct.
56 The argument is made in the essay "Myth Today" in *Mythologies* (New York, Vintage, 1993). See especially p.146.
57 Cf. Hannah Arendt, *The Human Condition* (Chicago, University of Chicago Press, 1958). Following classical precepts Arendt makes a major distinction in this book between labor (that which is not productive of durable things), work (which is) and action (the site of true ethical action). Killip's pictures in many ways reverse, refuse and scramble this logic.
58 Oxford, Oxford University Press, 1985, p. 279.
59 On capabilities and social justice see Amartya Sen, *Development as Freedom* (New York, Vintage, 1999) and, more recently, Martha C. Nussbaum, *Frontiers of Justice* (Cambridge, Harvard University Press, 2006).

in the accurate American parlance—in Aristotle they are denied citizenship). To reverse this process and, without idealization, to redeem work, not for itself but as a mode of revelation of consciousness that goes beyond the limits of inherited norms, necessitates the recovery of what is within work or in the act of working that allows the subject to 'come through it.'

In the *Pirelli* project this ambition is set in motion by focusing the bulk of the essay not on labor but on work as the concentrated interaction of mind, eye and hand. This focus is particularly evident in two great sequences of pictures (one near the beginning, the other the final dozen or so pictures in the essay). These pictures are clearly what the project is fully *about*. In them Killip pictures the acts of labor as *essentially* focused concentration. The essay as a whole offers a small dictionary of the modes of mental engagement in work, running the gamut from the heavily physical to the wholly cerebral. In these two sequences the line between labor, work and mind definitively dissolves.

Labor becomes mind and the picture is the portrait of consciousness as work.

That this touches on Killip's intent is suggested by one indicative fact.

Despite the essay nominally focusing on "work," with all that usually entails for the body as a complete entity, only three of the pictures show a body full-length, most—and most of the most compelling—are what Nadine Gordimer talking about David Goldblatt called 'sectional photographs.' Gordimer was referring to Goldblatt's occasional use of "sections" of the body as sometimes more revelatory of the sense of self than a full body photograph.[60] In the *Pirelli* pictures, Killip applies this logic, *but in reverse*. His "section"—his location of the "sense of self"—is not the body but the upper torso and head, and most emphatically of all, it is the look, the look (of concentration and involvement) that is addressed to the task. What is photographed then in these pictures is not work but the *subject working*. In this transference we move from the domination of work (the subject subsumed under work) to picturing the subject *through* work—we move, with all due allowance, from illustration to portraiture.

XV — But before we can tackle the implications of this last point we need to consider the question of method.

As his own account makes clear, what Killip had not bargained for when he agreed to the Pirelli contract was the character of the factory he was to photograph in. Rubber in its early stages of processing is extremely light-sensitive. The Pirelli factory therefore almost entirely excludes natural light. This immediately posed a technical problem. As he had moved from the formal portraits of the Manx period, Killip had developed a technique of handling natural-light photography that made taking 4x5 pictures as nearly unobtrusive as possible. It was natural for him to seek to use these same techniques in Pirelli. His account reveals the impossibility. The light levels were simply too low. After two months of determined but inconclusive effort (which came close to wrecking the project) he abandoned natural-light and going back to skills learned from his apprenticeship, re-thought his approach in terms of using artificial light.

This had considerable implications for the manner of the work. Rather than the scene or the moment caught almost on the fly, revelatory moments are now necessarily first seen and then as it were anticipated in their potential recurrence. The moment is stalked rather than captured. A greater element of conscious planning and pre-thought is now involved in each shot. And this changes too, at least to a degree, what and who is photographed.

The earlier pictures in the series—now largely destroyed (only one survives in the published series, (see page 36)) almost naturally inclined to inclusiveness. As in the *In Flagrante* pictures, the subject was seen in context of their lives—in this case work, the factory. Some of this sense of context survives in a few of the pictures in the essay, but most—and virtually all of the most compelling — reverse the process. Abandoning the wider context of the factory as a whole, they focus in on the subject at-work, sometimes in astonishing physical proximity, often with minimal contextual reference.

In these pictures the use of artificial light is key—for it helps in fact to solve a particular difficulty. An acute problem in all picturing is to get close enough to the heart of the matter so that what is produced is not merely an illustration of what-is but a revelation. It is the problem of finding ways of transcending the given, overcoming limits; of being able to use the act of picturing as a *mode of discovery* not simply of repetition. Far from it being a matter of the existent trumping the means of its representation, photography's struggle is perennially *with* these means; it is the struggle to overcome photography's chronic weaknesses and limitations (to make "the raw stuff of the real" come through *despite* the photograph), to make the existent *mean*. Were it not so then Sartre's famous comment about photographs—that in them the persons shown, though present, are so 'without existential posture'—would have no resonance. But this statement ties directly to our experience with photography, for rather than the existent being brought to us through the photograph, it as such disappears from us. (At the moments that Barthes insists on the presence of the existent within the photograph, he tells us equally of his all but failed search for his mother *within* her pictures).

Appearance does *not* (automatically) yield presence, let alone insight, knowledge or discovery. The struggle in photography is always that of working photography's properties and conventions (its possibilities) such that through it presence can appear. Artifice is one way of achieving this. All artifice is theatrical. Used with intelligence, the theatrical transforms 'what happens' into an event. Theatre has no other meaning than this. In photography the form of the encounter between subject and photographer and the use of light to dramatize this relation are two of the simplest ways in which this can be effected.

In the Pirelli pictures as we have seen, distance, or even better the gradations of proximate, is the vehicle of the former, the use of artificial light the agent of the latter. Lighting theatricalizes the moment. In part it does this by erasing undesired context and allowing focus on precisely those moments of the event taking place that the photographer wishes one to focus on. Second it creates an ambient focus. In terms of the first, this is done so powerfully that in a number of the pictures—(see particularly the final four images, pages 57–61)—that background and context disappear almost completely. We are left then wholly with the illuminated subject. At the same time, the heightened moments of light/dark contrast throw into relief the central thematic issues of these pictures. Essentially the mobility, fluidity and plasticity of the subject counterpoised to the apparatus with

60 Nadine Gordimer, "Sudden Life, Never Seen or Suspected Before: David Goldblatt's Photographs" in David Goldblatt, *Fifty-One Years* (Barcelona, MACBA, 2001), p. 436.

which they are engaged and through which work is being performed. In the more contextual pictures, most of which come at the beginning of the essay, this is sometimes to the point where the subject all but disappears in the bulk of the machinery (see pages 16–17). More usually, and more powerfully, the heightened visual contrast tells. Making the subject emerges from the picture and thus to make it abundantly clear that it is this subject with which the photograph is primarily concerned.

The subjects emergence from ambient darkness is a conceit, but one that plays into Killip's goals—for these are precisely pictures of the emergence of subject's "out" of work, out of the ambience of work, actual and conceptual or categorical. This is the subject who emerges not as a figure in a documentary but as a self-sufficient subject, as essentially a portrait.

XVI — Of course, in so far as a certain credibility that "this is" has to be maintained, that these pictures must be seen to deal with the facts of the situation, then the theatrical quality of their production has to be downplayed. Work must be performed to keep the sense of the artificial in check. The viewer cannot be allowed to presume that these are fictional moments. There can be essentially no sense of "moments staged" (even if, as we will see much rhetoric, and much staging, actually comes into play). It is telling in this respect that in only one picture (see page 38) is the presence of photographer overtly acknowledged. In most, and even in the most intimate pictures, where the physical distance is extraordinarily close for male-on-male encounters, the fiction of the "invisible" photographer is maintained. It is this induced (but nonetheless astonishing) lack of acknowledgment of Killip's presence (and with it, conversely, the absorption of the subject in the task) that sustains the "documentary" quality of these pictures, their (necessary) sense of being, at least to a degree, observed records of factual events.

The tension between these two moments—the factual and the theatrical or the fictive—is acute and is central to the work the pictures achieve. The felt veracity of the moment, the event, holds us. Actuality yields credibility. And yet the sense that the photograph *reveals* actuality—which is after all the most fundamental and therefore the most *obvious* fact about photography—at least at the level of appearance, can be given a deeper twist if we add an insight offered by the novelist Harry Mulisch. In his recent novel *Siegfried,* Mulisch has the character Herter (also a writer) propose that fiction or artistic phantasy is sometimes less that which must be understood than "something *with which* you understand."[61] Although this observation is something that is, at least ritually, acknowledged in criticism (the value we place on all art ultimately rests here: this differently, is Nelson Goodman's insight as we find it expressed in some beautiful lines in *Languages of Art*[62] in the case of photography, the relation between phantasy and truth has always been ambiguous, not least because psychologically the ambition for "truth" or its aspiration (as the revelation of essence) displaces understanding—to the point where (*pace* Barthes) it will be denied that the photograph can offer understanding. Mulisch's observation is a strong reminder that art is that fictive-yet-real construct that is created not just *from* understanding but *in order to* understand. The fictive is the tool by which the real may be grasped.

Understood correctly this insight restores to photography its critical role. Work is that *with which* you do something, and the vehicle of that doing is the moment of phantasy, of imaginative comprehension, that the photographer adds in the moment of taking the photograph. Photographs offer understanding—are tools of comprehension—because they are created *from* understanding, and because in turn they exemplify this understanding in a form that makes it available to the viewer, i.e., makes of the work something (an event) *through which* or by which the viewer understands a situation or a subject.

In the Pirelli photographs the pictures are the frames through which we look at understanding of work. Everything hangs on the intimacy involved, both with work and with the subjects pictured.

Killip has spoken movingly of distance in relation to David Goldblatt's photography: "Distance, the correct distance, is the great pivotal key to gauging a photographer's worth and I have very strong feelings about this, the greatest gift his photography possesses. The distance, from camera to subject, has to be unquestionably the correct distance, which I cannot explain, only recognize. It is a variable that lies somewhere between inches and miles. From photograph to photograph it changes as the subject, experience and intuition dictate."[63] This applies, clearly, to his own photography as well. In the Pirelli pictures distance articulated as proximity, even intimacy, is what propels these pictures out of the realm of the documentary (the recording of a figure, at work) and into the realm of portraiture. There are pictures in the Pirelli series that "document"—not literally but in the sense of providing a circumstantial documentation of processes and content. But these pictures are there to provide almost the alibi for what is central to the essay—the thirty to forty pictures that constitute, taken together, a complex group portrait of the Pirelli workforce, and which singly provide us with a series of individual studies of the subject-in-work.

XVII — Are these pictures then portraits? Should we see this series not as a documentation of work (it is anyway not this) but as an extraordinary extended and sequential group portrait? It would seem so—with one significant qualifier.

The appellation "portrait" is traditionally applied to a picture of someone who is *named* in it, but the wider and more accurate definition is that a portrait is a picture in which its occasion—the subject who (for whatever reason) impelled it—has a presence *relative* to that of the picture itself. Indeed, the portrait is a celebration of this fact—a representation that acknowledges the presence of the subject who occasioned it. The portrait behaves then as if what we recorded Barthes saying earlier of the photograph—that 'in it the event is never transcended for the sake of something else'—were true of all picturing. It is this Barthesian quality (which we take as true *enough*) which explains the affinity between photography and the portrait. Precisely this was what popular imagination quickly discerned as photography's potential—all the more so that this (desired) quality was so much more accessible through the photograph than it had been through painting. Although the aristocratic dominance of portraiture is less than is often supposed, particularly in the century before photography's invention, nonetheless the particular contribution that photography has given to the portrait in the C20th—the anonymous or essentially anonymous portrait—is sufficiently rare in the history of painting[64] that its high points—Hogarth's pictures of his servants for example, or most astoundingly, the five surviving pictures from the series of portraits of the insane that Gericault painted in the early 1820s—come quickly to mind.

61 Trans. Paul Vincent (NY, Viking Penguin, 2003) p. 13
62 Indianapolis, Bobbs-Merrill, 1968, see especially p. 258
63 Chris Killip, "A Photographers Life" in David Goldblatt, *Fifty-One Years* (Barcelona, MACBA, 2001), p. 427.
64 It is much more common in the history of drawing.

It was inevitable that photography would develop towards the anonymous portrait. Three factors made it so. The first flowed from photography's sense of curiosity concerning the world. The desire to depict other subjects, initially as types, objects of curiosity, was endemic from the start. But, and this is the second factor, in such photography it is impossible—despite the clear attempts of many early photographers—to wholly suppress being as such. The truth that photography is made to discover here, even against its own inclinations, is that the force of the subject cannot be entirely contained within the conventions of picturing types. A beautiful example of this, and one that neatly parallels Gericault's pictures, are the pictures of the *Inmates of the Surrey County Asylum* in England taken in the early 1850s.[65] These figures address the camera with a naivety that shatters the conventions of the anthropological or sociological look of the type. This has the immediate effect of rendering them as distinct consciousnesses, no less individual, no less distinct as subjects than the *personnages* who sat for Nadar.

It is this assertion of the subject *within* the type (the subject *excessive* to type) that begins to distinguish the anonymous portrait and which brings us to the third factor which motivates it and that is politics. The turning point is probably Sander. Sander's is a nineteenth-century project of typology and classification carried out with a twentieth-century sensibility. In it the project of typology breaks down under the weight—the presence—of asserted subjectivity. It is as if Sander cannot help but photograph the subject "beneath" or "beyond" the type he wishes to record—to depict being not in terms of type but being "such as it is."[66] Opposition to Sander's project by the Nazis was not (as is usually said) because his project showed the "wrong" types but because the Nazis recognized that, in effect, his portraits destroyed the notion of type altogether, i.e. they destroyed the notion of simply *naming* subjectivity.[67]

It is difficult to over-estimate the political significance of this point.

We have already seen that one way that photography anticipates is by bringing subjects to visibility as such, i.e., as subjects *for* politics. The force of the anonymous portrait is that it brings being into play not as under an already determined classification but in its potentiality—being "such that it is." The Pirelli pictures are *reminders* of being-in-work, even of being-through-work: they are in Killip's words, "a dissection, an examination, an uncovering and ultimately a confrontation: a showdown," both with work and what we assume of it vis-à-vis those who work. "To show work, revealed through those who do it, who are affected by it; made and unmade by it," proves in the end to open to what being is today, namely not to that which is determined by given, or above all *ascribed,* characteristics (whose intent is always to *place* being) but as that which happens, as the philosopher Gianni Vattimo puts it, when we 'historicize ourselves,' when we act. Being, in this sense, "is" not. It happens—it occurs—in the immanent event of our acting.[68]

XVIII — Yet it is axiomatic to the tradition of portraiture that the portrait is a picture of the subject being and not doing. There are technical reasons for this of course—most obviously the requirement (for painting) of poses held in duration. But there are equally ideological, even moral reasons. It is in the end for our being even more than for our actions (which so often do not bear close scrutiny) that we wish to be known. The portrait therefore is traditionally a pause in action: a presentation of being. The continuity of life is interrupted in order that being can be recorded. It is this "stop" in the life of the other that, when successfully transformed into an image, causes in us a parallel "stop" in our continuity—the opening to the moment in which reflection on being can happen.

But in the Pirelli portraits there is apparently no "stop" of this kind. In only one picture is the subject fully stopped or held in this sense. In two more they are caught in mid-process. In all the others the flow and the rhythm of work continues. These are then, almost uniquely, portraits that happen in-, or better *through-,* acting. Traditionally, the portrait is the *result* of actions that the subject had performed; it is recognition of the success of past action coded into a picture that celebrates the being (the presence) of he who performed the act. Killip's pictures are significant transformations of this theme. In the *Pirelli* pictures the portrait emerges through the picturing of the act.

What does this mean, and especially for the politics of this project?

The portrait is always a *staging* of being. In a good portrait appearance is animated by the tension between the need to hold onto the *fact* of the subject's existence beyond or through the act of representing them, while in the same moment deploying appearance (representation) not to negate or to subsume this existence (such that the picture makes of them a type, a figure, a personification—*representation without existence*), but to achieve the exact opposite, that is through the *intensification* of the act of depiction to *bring the existence* of the subject, in their presencing, to visibility. In a number of pictures in the Pirelli essay Killip achieves this goal, but he also perhaps does something else.

Being is a consequence of how we act—of how we are *able* to act. Politics today, meaning a proto-democratic politics, is not a politics based around seeking to realize already given entities or categories, but a politics that arises from an understanding of this relationship between being, acting and capabilities—which means potentialities. Politics, a democratic politics, concerns realizing the capabilities of all. It is opposed to those elitisms, which in one way or another, have pre-determined identities and therefore capabilities and therefore also determined who can, and who cannot, have freedom of action. Beginning from the late 70's, and acting in the interests of an emboldened drive to inequality, there has been a sustained economic attack on the democratic rights of what used to be called the "working-class"—above all by the removal of work and with it then the ability to effectively exclude those pushed out from labor from anything but marginal existence in relation to society. Part of that attack—by no means the least of it—has been the denigration of labor as such, and through this, crucially, a denigration of those-who-labor.

Killip, in his pictures in the 1970s and 80s, was the foremost chronicler of exclusion in Britain, perhaps in Europe. These photographs, far from standing as historical documents to a past time, are today of increasing relevance. What they picture, as the riots in France this year showed, is now coming to pass across a vastly wider geographical area than almost anyone would have thought possible twenty years ago.

The *Pirelli* portraits, by contrast, are pictures of work, but they too are no longer historical—though their anticipation is not at all where we might have supposed it. Looked at through the lens of the already rapidly fading rhetoric and history of "industry", "the working class" and "labor", the *Pirelli* pictures are simply anachronistic. But they are not so, the work they explore is not the work we assumed. It is not as historic as it first seemed. The project has already lived past its origin. What it speaks to now is the subjectivity and politics that emerge from grasping "being such as it is." That subjectivity is not fanciful. It is transitive. Hence these pictures have much more to say to the future than might have been supposed. *In Flagrante* was the exploration of exclusion from a perspective that was irreducible to the given orthodoxy. The Pirelli pictures are an exploration of work and of the being that emerges through work and in work which is irreducible to all that we thought work was and to all that we thought those-who-work are. These pictures cannot be experienced as things already in the world because although grounded in an actuality (but one different from what we suppose) they present us, almost secretly, with a new perspective on the world—an inaugural grasp on an understanding of being that is no longer tied to the apron strings of history, but instead defined in terms of our capacities and our potentialities.

65 The photographs were taken by a Dr Hugh Welch Diamond, ostensibly as physiognomic portraits of disorders.
66 The reference is to Giorgio Agamben. See his *The Coming Community* (Minneapolis, University of Minnesota Press, 1995.
67 Badiou, op. cit., has a useful line: 'What matters here is the general principle. Evil in this case is to want, at all costs, and under conditions of a truth, to force the naming of an unnamable. Such, exactly, is the principle of disaster" (p. 86).
68 Gianni Vattimo, *The End of Modernity* (Baltimore, John Hopkins, 1991) p. 3.

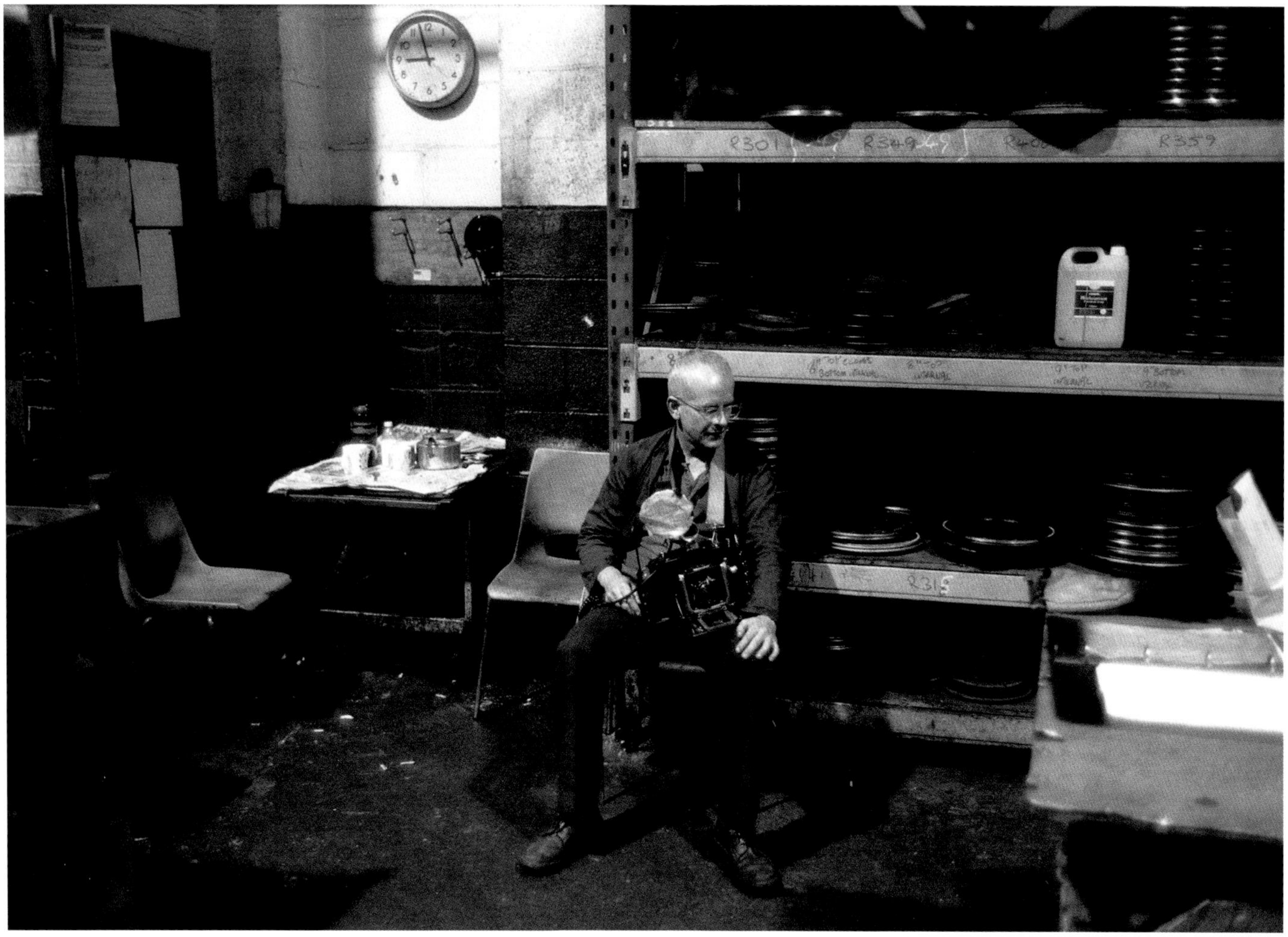

Portrait of Chris Killip in the Pirelli factory, August 1989, Keith Pattison

Acknowledgements and thanks

My grateful thanks go to the entire workforce of Pirelli at Burton on Trent for their generosity and forbearance during the making of these photographs.

I would also very much like to acknowledge the unconditional support I received from the instigator of this project, the late Tom Northey of Pirelli UK and Sandro Veronese, the former managing director of Pirelli UK. Also to Mark Haworth-Booth of the Victoria and Albert Museum for agreeing to exhibit the photographs, sight unseen and finally to Keith Pattison for all his help.

First published in 2006
Second edition 2015

Edited by Ute Eskildsen
Book design by Victor Balko / Steidl Design
Tritone separations by Steidl's digital darkroom

Production and printing: Steidl, Göttingen

Steidl
Düstere Str. 4
37073 Göttingen
Germany

Phone +49 551 49 60 60
Fax +49 551 49 60 649
mail@steidl.de

steidl.de

ISBN 978-3-86930-961-3
Printed in Germany by Steidl